THE AMERICAN DEMAGOGUE

THE AMERICAN DEMAGOGUE

Donald Trump in the Presidency of the United States of America

MOLEFI KETE ASANTE

Universal Write Publications LLC

THE AMERICAN DEMAGOGUE: ***Donald Trump in the Presidency of the United States of America***

Book Designer: AuthorSupport.com

For information:
Website at www.UniversalWrite.com and www.UWPBooks.com
Publisher: Universal Write Publications LLC

Mailing/Submissions
Universal Write Publications LLC
421 8th Avenue, Suite 86
New York, NY 10001-9998

ISBN-13: 978-0-9825327-7-5
ISBN-10: 0-9825327-7-6

Arthur and Lillie Smith,
whose descendants continue to cover the earth.

Acknowledgements

I want to acknowledge my publisher, Ayo Sekai, whose dedication to academic publishing has become one of the heroic tests of resilience in a world full of competition for recognition of good works. She has been devoted to making Universal Write a company of skilled editors and authors who will change the nature of Afrocentric scholarship. May this book, and many others, be added to this line in the interest of unfettered rational thought.

CONTENTS

PREFACE

Demagogues tend to be the same across generations and countries and in this book I attempt to show how Donald Trump in any guise is no more than a petty demagogue whose acclaimed narcissism and sociopathology are similar to that of other demagogues. Clearly the rise of Trump to the presidency of the United States signals the radical abandonment by the American electorate of the long arc of progress that had become the hallmark of every president since Franklin Roosevelt, Democrat or Republican. Trump has taken up the politics of fear, the rhetoric of intimidation, and the temperament of a bully in his relationships with members of his own government, his opponents, and other world leaders.

No person has ever attained the presidency with such anti-institutional bias as Donald Trump. His attack on the security institutions, FBI and CIA specifically, was crass and mean-spirited and his attempt to shut down the media institutions that did not support his presidency was frighteningly similar to the way the German *Regenpropagandaministerium* under Hitler and Goebbels attempted to control information during the Third Reich.

Trump like Hitler took his rhetoric of fear to the social media of the day in order to persuade the millions who could listen to him or read him without the important critiques of the experienced media commentators. Like other demagogues Trump understood that the control of the security and media institutions was central to his ability to dominate the imagination of a nation. Trump would do this by employing the platform given to him by the Alt-Right and the Fox News Agency and Breitbart to spread his version of fear, hatred, enemy opposition, and messianic delivery. It is necessary for those who understand the decline of civility, the abhorrence of reason, and the whipping up of mean-spirited emotions to speak truth to authority and to challenge hatred with concern and care. Demagogues do not get better or more civil; they demand more attention, power, and control. Speaking against such decline in society is what the conscious intellectual must do.

When Trump addressed his rallies during the election campaign of 2015 for the presidency he had the instrumental support of the Russians who worked to elect him president of the United States. All of these rallies had been followed by his bold and racist announcement of candidacy at Trump Tower on June 16, 2015 when he said, "When Mexico sends its people, they're not sending their best. They're not sending you. They're not sending you. They're sending people that have lots of problems, and they're bringing those problems with us. They're bringing drugs. They're bringing crime. They're rapists. And some, I assume, are good people." This is widely believed to have been Trump's signal to what would become is political base, the white racist population that hated the integrated, diverse, and multi-racial country the United States had become. Regardless to whatever he would say in the coming weeks and months this quote would never be corrected, changed, or modified by Trump. It would follow him as surely as ants would follow the path to sugar. He was not just stuck with the words; he wanted the words to be his so that he could use them as bait to bring in those who felt they needed, after Barack Obama, a white savior. But of course, saviors do not come without cost to national morality, the bruising of core values, and the demise of free speech.

What Trump meant was that he was the One, when he said, "Now, our country needs— our country needs a truly great leader, and we need a truly great leader now. We need a leader that wrote "The Art of the Deal." We need a leader that can bring back our jobs, can bring back our manufacturing, can bring back our military, can take care of our vets. Our vets have been abandoned. And we also need a cheerleader." Most truly great leaders do not announce their greatness, it is bestowed upon them by history. Only the truly fearful, the insecure, the jealous, the bitter, and the misguided resort to bluster, bombast, and the penchant for control of all institutions. By exploiting the weaknesses of those surrounding the "Leader" Trump is able to find obsequiousness in the obvious suppliants who seek closeness to power. Those who have vulnerabilities, that is, those who are threatened in any intellectual or moral capacity become the pallets upon which the demagogue lays his trust because they have been considered not-smart, dirty, easily tempted to corruption, or incompetent. Rising to work in the interest of the Leader they become cadre of promoters who vow to feed the ego of the demagogue. They will often remain around so long as they are subservient, praise the most stupid actions of the demagogue, and do not feel frustrated by his personal insults. To the degree that the demagogue's base become like the demagogue they neither have integrity nor character.

I have written this book because I have faith in the inexorable nature of people power and I believe that it is possible to stop "he-who-would-become-the-Leader" by the rising up of the masses that refuse to be dictated to by any person who rides roughshod over democratic processes. As an African American I have had to believe in the ordinary people, attached to laws, and to a certain degree of morality and common interest in peaceful coexistence that would sustain the movement toward justice by underwriting a national commitment to the equality of all people.

Introduction

Perhaps no American presidential campaign or presidency has been contorted with as much anti-American rhetoric as the campaign of 2015 and the presidency of Donald Trump beginning in 2016. I mean I do not recall a candidate who has taken on the memes, core values, representative institutions, and optimism that are usually associated with the American nation and turned them into such negations of promise, as has Donald Trump. There is no question that the Trump experience will be seen as a pivotal project where the deep continuity of a nation was fractured by a wildly unique and wily politician who capitalized on his television celebrity to bend the nation toward moral deformity for political purposes. What one wonders is whether or not Donald Trump knew that humanity was dirtier, meaner, and more evil than others knew?

In many ways Trump is the anti-American, the twisted and distorted version of what an American was thought to be. Indeed, if one thought of Americans as being progressive in their relationship to the pollution of the earth, then Trump says pollution is good. If you thought of Americans as proactive against racism and racial injustice, then Trump proves that some on the side of racism are good. If you understood Americans to be opened to the weak, the hungry, and the poor, then Trump proves that he does not have the same compassion demonstrated by other presidents, either Republicans or Democrats. One can select any set of American institutions, either security or media, for example and in each case Donald Trump has demonstrated an easy ability to bash those institutions or organizations he views as opponents to his brand of politics.

Consequently my objective in this book, given my background as a scholar in the field of communication, and as an Africologist with an eye toward the negative images, symbols, language, and references used by Trump to denounce his predecessors, especially President Barack Obama, was to reveal the dirty underbelly of an American Demagogue who peddles lies as his political bargaining chips.

Many years ago one of my former students at the University of Buffalo, Erika Vora, wrote a dissertation on the subject of diffusion of information using some of the rhetorical addresses of Adolf Hitler as the basic documents. What I learned from reading

Vora's study was that the demagogue could effectively turn the ordinary citizen into an angry, irrational, and violent votarist simply by lies, appeals to fear, and creating tactics that suggested one would lose a job, a spouse, or a child by allowing "foreigners" to enter the country. Trump pulled out every instrument of persuasion at his disposal to obliterate rational thinking from the very first day of his presidency when he and his people claimed that they had a larger crowd at Trump's inauguration than was at Barack Obama's inauguration. It was a large lie, an in-your-face bold lie that caught most people off guard. During the next few days Trump established a pattern of dominating the news cycle with a tweet a day, sometimes several tweets a day, and then watching the commentators, news anchors and general population discuss his latest lie. I recall that Hitler had tried to do a similar thing with radio; indeed, his was the first real radio demagoguery. Trump resorted to Twitter, using it to create a topic line for the day, thus truncating any other line that might interfere with his visibility in the media.

Democracies are governments by rational argument. The United States of America has often been described as "the leader of the free world" where government by talk, not by fear, bullying, or lies, dominates the public sphere. When I felt that this tradition was under threat I decided it was necessary to revisit the making of a demagogue. One could have talked in the past about American politicians being racist, ignorant, misogynist, and segregationists but no one had reached the height Trump reached with a ranting, ignorant, lying behavior simply by appealing to the worst instincts of human beings. According to Trump, if you felt that your opponents were not human, then you could "knock the crap" out of them. If you felt that Mexicans would not be good judges then you would try to avoid Mexican judges.

In this book I intend to examine how fear is used to create a condition of radical anti-immigration and anti-foreign feelings. But this is not enough to make a demagogue; it was necessary for Trump to also tackle the most appreciated, loved, and respected individuals of the society. By demeaning their intelligence, downplaying their achievements, and damaging their reputations with his lies he recreated himself in the image of singularly unique and special leader who could correct the deeds of the past. Trump would challenge the popular Obama by continuing the assault on his origins, being coy about whether he would admit now that all evidence as in that Obama was born in Hawaii. He had circulated the lie prior to his election and those who believed him, so gullible were they and so needy for a savior, accepted the lie as if Trump had been at Fatima and had seen something that no one else could have seen. The energy that he put into attacking Obama was the kind of venom many black people have experienced in America. It was personal, gloomy, disturbing in its vileness, and carried with it an existential threat to the name and person of Obama. Many people thought that Trump's attitude toward Obama's Presidency bordered on a hateful jealousy of Obama. Nothing President Obama said or did could be entertained by Trump; his animus was clouded by a small ego writ large by insisting that he was someone other than who he was: a Queen's New York man who inherited millions from his father to begin his career. He would have cabinet officers vow to take down all Obama-era reforms and regulations, almost gutting the environmental protection agency by making it an agency that protected, not the people, but the large corporations. Health care was tangled into a web where individuals lost their ability to survive in times of crisis; Trump's attitude added to the uncertainty and existential threat that many sick people felt.

Although Trump peddled fear; he was mortally frightened by the possibility of losing the election to Hillary Clinton, a woman. She was the most qualified person to ever run for the office, having served as First Lady for eight years, as senator, and as secretary of state, with a law degree and achievements for the public good stretching back to her college days as a demonstrator for civil rights. Like all politicians she had her own set of problems and prejudices, but no one would say that her deeds were anywhere in the category of Trump's moral meanness and unhinged mendacity. Many of his actions showed that he disliked women, thought of them as less than men, and viewed them as objects useful only for his personal pleasure. Almost every woman who came into his orbit had to submit to his male chauvinism, deeply weakened ego, and ruthless patriarchy. He proved to be a fearful man and he bluffed his way to the presidency by appealing to latent racist proclivities among the least educated Americans.

There have been and there will always be debates about the true nature of democracy. The idea in Athens was that all citizens were equal although we know that Athens had its own problems. Those who had property were considered the true citizens and their voices were far more powerful than those who were without wealth. Rome would continue the process of developing democracy with an eye toward supporting the principle that wealth gave one more freedom. It would be centuries before the Founders of the American nation challenged the British Empire with their notion of individual liberty. Each person's house was to be his castle and there were to be no person above the law. Indeed the meetings in Philadelphia that established the founding documents, after the demise of the decade or so old Confederacy that had been set up after the Declaration of Independence, sought to enshrine the principles of a dynamic democracy where human being would be equal under the law. The Constitution, flawed as it was by the enslavement of Africans, the shunting of the Native Nations, and the invisibility of women in the law, held out the possibility of equality. Indeed this possibility is what morphed into the hope and optimism seen by those who could neither read nor vote; their sight was insight and their knowledge of what was in the Constitution was only from hearsay and rumor. Upon these thin threads dangling from the Constitution millions of men and women tied their hopes. What my ancestors could not read they could imagine, and to them there would be a time when they would the Constitution would fulfill its promises. I grew up with the feeling that we would eventually be able to agitate for our rights under the law.

Our struggle for human rights, often translated as Civil Rights against the legislated segregation and racist practices of the nation, helped to create the platform for numerous other movements for freedom. This is why we say that the long African fight for freedom has been America's true journey of liberation. Freedom, therefore, was defined by our struggle to throw off bondage. In many ways this was different from the liberty sought by white settlers protesting taxation without representation. Our struggle was for freedom from enslavement.

Civility can be fleeting given the rhetoric of a lying demagogue. Even individuals who are normally civil toward their fellows discover in the rhetoric of the demagogue an evil path toward community dissension and often they stride toward society's breakdown thinking that they are operating in their own best interest when in fact they are simply falling into a dangerous trap laid by the demagogue for his own political interest.

The aim of laws is to regulate human activity for the interest of the whole but even laws cannot prevent citizens who are full of racial, class or religious animosity toward their neighbors from listening to the demagogue who seeks to divide people. This is different from dividing people on the basis of arguments against or for a particular policy. One cannot claim to be for democracy while at the same time distorting the meaning of democracy by shaping it in the image of race, religion or class.

Inherent in democracy, especially when the institutions of a democratic society work, is the ability of the people to throw off the demagogue. This is the promise we have inherited as a nation through the sweat and blood of many movements in the last century such as the Civil Rights Movement, The Students' Movement, the Women's Movement, the Black Power Movement and so forth. The clearest assault on irrationality is rationality; the key to unlocking shady operators in mendacious words and actions is confrontation sharply and often. A demagogue feeds off of the shouts and applause of the crowd like a vulture off of dead meat. In this case the demagogue throws the dead meat to the audiences and when they respond, he in turn, responds to their applause. In reality, both the demagogue who has thrown the wordy meat and those who have received it share in killing any form of rational dialogue. They form one long monologue of self-glorification.

This book is not complete because the presidency of Donald Trump has not ended. However, I believe that it is necessary to record how we got to this place in the American democracy. I mean this place where we are no longer the beacons of hope for the millions who are destitute in the world. How did we get to this place where we are willing to roll back the clock on health care, environmental protection, disease control, and racial integration and diversity? What I have tried to demonstrate using the best angles on data that I can muster is that the resilience of the American democracy throughout history has always been able to smother the demagogue.

Chapter I

The Demagogue

As a student during the 1960s I had a teacher in communication whose name was Friedrich Casmir, we called him simply Fred L. Casmir.[1] By the time I enrolled in Pepperdine University for my Masters in Communication Fred Casmir had already made a name for himself as teacher and preacher. He had been in Hitler's Youth Corps and after the defeat of the Nazis had entered Ohio State University and got his PhD. He traveled the United States giving a talk called from "From the Swastika to the Cross."[2]

It was from Fred Casmir that I learned about the dangers of demagoguery although I never quite believed at the time that he had completely purged himself of the effects of being in a society where the supreme leader wanted others to see him, as many did, as a demi-god with unimpeachable qualities. It would be later after the Black Power Movement, Casmir's losing campaign to be in the House of Representatives, many conversations, and Casmir's tremendous faith in my career goals that I began to see, and he began to see, that we could be friends.

I met for the first time Leni Riefenstahl's film *Triumph of the Will* which was considered one of the most powerful propaganda films ever made at that time. The film deeply etched itself into my memory and reappeared in my imagination every time I learned something about the fickleness of rationality. I have thought much of the willingness of the Germans of the Third Reich to give up their own rationality to the rambunctious and determined cadre of Aryan supremacists who surrounded Adolf Hitler.[3] Many were ordinary citizens trying to eke out a living in an unstable economy.

Wearing their brown shirts and shouting cruel and terroristic threats at Jews, Roma, and gays, the Nazi votarists led by extremist thinking, and the rhetoric of Adolf Hitler, dominated the airwaves and the media attention in Germany. They became Hitler's oratorical barnstormers who set the nation on fire as it waited for the words of der Fuhrer.

What propelled Hitler to the leadership of Germany? Given the fact that Hitler openly preached anti-Semitism and believed in the superiority of the German people, why would the people see him as a leader? For him, there was nothing wrong with the German society except it had opened itself up to an "invasion" of non-German people. The idea was to create and maintain a purely German state with people who had German blood.[4] All demagogues like using biology to separate their people from the aliens. Once they have identified the enemy as other then they are ready to destroy the other.

The Normalization of Demagoguery

I think that we should remember that Hitler and the Nazi Party lost the election of 1932 but President Paul von Hindenburg appointed him chancellor in January 1933. A few months later in March the Reichstag passed the Enabling Act of 1933 that sealed Hitler's rise to power. It had been a mere ten years since the Munich Beer Hall putsch that introduced Hitler's bombastic racist brand of oratory to the German people.

The demagogue believes that he has a right to rule because he is smarter than other people. In a twisted way he sees himself not as Plato's philosopher king but as the natural leader because of his supposed higher intelligence. In Trump's case there is little evidence that he has a liberal education or a wide understanding of the contemporary world. To attend school or college is not the same as being educated. Trump neither reads deeply nor is he willing to be taught by those who do read seriously. He is like a fidgety child who responds only to commands barked by someone he believes is more powerful than he is.

Hitler was the one Nazi Party speaker who could arouse the raw emotions of the crowd again and again. Finally he had threatened to resign from the party if his colleagues did not make him leader. He believed that he was the most gifted and the best suited to be leader and consequently the people made him their leader.[5]

Donald Trump's desire for power came forth out of revenge, something that he has felt most of his life, probably even when his father placed him in military school to "teach" him discipline. No one can tell what really happened in that school. Did the young bully meet other bullies who were bigger and stronger?

It does seem that this experience gave him a deference and reverence for those in military positions. He came to love the title "General" and I believe would have preferred for those around him to see him as the General Great. This is why generals with names such as Flynn, McMasters, Mattis, and Kelly were prominent in his first government.

All one has to do is to examine the life and career of the most prominent European demagogue of the 20th century to get a glimpse of what an American demagogue would look like in the 21st century.

I have no evidence that Trump studied the speeches of Hitler, but I have a powerful suspicion that Hitler was a figure he admired in German history, and being impressed by Hitlerian rhetoric may have been a part of his family's attachment to German martial heritage. Using demagogic language and wild conspiracy theories Hitler spoke to his political base, people who felt left behind economically and culturally, in an effort to "wake them up" to the dangers that he saw surrounding them.

The Danger in the Other

Trump sees an America that is overrun by immigrants. In his judgment the nation's southern border is so porous that the country will soon be filled with undocumented aliens. He has been able to give this faded picture to his political base in towns where his followers were the last immigrant families and they only see contemporary immigrants on the television. They have often drunk the lemonade, so to speak, or hung the picture he has given them of immigrants on the walls of their brain.

Der Fuhrer showed his mass audiences in Germany the same broken picture. For Hitler it was the "yoke of the Jews and Communists" and for Donald Trump it was Muslims, Mexicans, Africans, Asians, and Haitians who were dangerous for the American nation. Hitler preference for the Aryan and Nordic types can now be seen in Trump's preference for the Nordic and Aryan people. He infamously said, "Why can't we have more Norwegians" after he had decried the immigration of Africans and Haitians from "shithole countries."[6]

When Trump announced his run for the presidency he indicated in that speech his beliefs about people who were not like him. He declared, "When Mexico sends its people, they're not sending their best. They're not sending you. They're not sending you. They're sending people that have lots of problems, and they're bringing those problems with us. They're bringing drugs. They're bringing crime. They're rapists. And some, I assume, are good people."

Straight out of the classical demagogue's play book Trump chose an "other" to begin his campaign. His was not a speech of promise, hope, progress, and vision but a dark and stained view of petty politics using the model of the demagogue. Attacking the Mexicans seemed particularly easy and divisive since the Mexican population is set to grow rapidly in numbers and Trump's sense was that to tap into the animosity that many whites have against "colored" immigrants would gain voter approval among white populations.

I have yet to hear that Trump is calling for a white empire that would "rule the world for a 1000 years" which is the preferred number for demagogues such as Ian Smith of Rhodesia and Hitler of the Nazi regime. In Africa the last true demagogue was not Robert Mugabe but Ian Smith who made the appeal to the white Rhodesians for a regime that would rule for "a thousand years." However, what Trump has done instead is to rile up a minority of whites who are solidly in his camp, and against immigrants, and who would insist that the door be shut on what he has called "chained immigration" in a very hyperbolic way. What is this *chain migration* except the most humane form of immigration? It is precisely the path taken by many European families; they came and they sent for their mothers, fathers, nieces and nephews, and sometimes their grandparents. Why should this door now be closed to contemporary immigrants? Are we sensing something ominous about the nature of Trump's rhetoric? Is the fear of others so prominent in his mind and among his political base that we are seeing evidence of the breakdown of the American ideal?

Whenever you insert the seed of racism in the immigration policy of a nation, especially one built on the basis of people from other nations, you are creating a dangerous and explosive situation because it pits those identified as citizens against non-citizens, whites against Mexicans, African, etc.

Sometimes it is useful to have a good American history lesson. But perhaps we should start with a history of world democracies. Here we would see so clearly that democracies do not die; they are murdered when the people lose sight of the democratic vision. Democracy is a powerful idea but it is also difficult when you have agitators willing to substitute the rule of the rich or the rule of race for the rule of the people. Another reason reading is important not just for the leaders but for the masses is to be able to make wise decisions. Demagogues insist that people cannot rule themselves without leadership of the "smart" ones with the "really good genes." Here a nation descends swiftly into bitter and sometimes warring factions led by a demagogic tyrant. History teaches that there is always only one end to that road: the demise of the nation and the end of the tyrant.

Michael Wolff says that Trump had the obsession of firing people, for example, Rosenstein, Comey, Mueller, Wray, Sessions or whoever is the person of the day that he feels threatened by.[7] This fascination with ending the career of opponents in a crushing way is the spirit of the tyrant. Of course, he has all of the flaws of a petty leader: a desired patrimonialism, paranoia, profligacy, populism and jealousy of previous leaders.

The demagogue constantly goes to his political base for rallies where emotional energy is always kicked up by being against something or someone. Trump has a depressingly large number of targets that completely drive his audiences nuts with yells and screams when he mentions their names. His attacks on anyone that appears to threaten the way he perceives himself are always couched as his way of defending not himself but his political followers. The rallies of the right are always full of venom; Trump stirs up the most base emotions and then he leaves the people to see conspiracies everywhere although they have nothing of substance to which to point. Mexicans are not criminal and Africans do not come from "shithouse" countries. Stupidity reigns only so long as truth is not revealed.

The Only Smart One

Trump tells his political base that leaders before him were dumb, stupid, and political fools, who got swindled by foreign leaders and nations such as China and Mexico whose leaders were smarter than American leaders. During his rise to the presidency Trump attacked all of his opponents and even the progressive actions that had been achieved by presidents before him. He showed his personal angst at Obama's liberal and humanistic legacy and vowed to take down all of the executive actions of Obama. His speeches attacked trade agreements, the Iran agreement, NAFTA, and the Pacific-Asian Partnership, the Paris Accords on Climate, and anything that Obama was in support.

"I am really smart," the demagogue says to his emotional followers who often look to his business dealings as evidence of how smart he really is. But a person who inherited millions and works to maintain those millions, maybe to increase the amounts, is not necessarily demonstrating evidence of intellect, but could be demonstrating the evidence of greed, stealth, theft, and bluster. Money cannot be equated with wisdom nor intellect. The Odu Ifa says that if you have money but not character the money does not belong to you. Ifa wants only character,

Most Bullies are Weak

Donald Trump like most bullies is a fake. However, it is his insecurity based on the fact that he is not as smart as those who read rather than boast that propels him into accommodation with demagoguery. Why would an American president want to ask the military to provide him with a martial parade to show off America's weapons? Have we come to the end of the American era? How did a demagogue interested in his own glory, not that of his country, ever rise in a democracy? These questions sit at the door to understanding the assertion of the demagogue to gather power to himself.

The Micro-Aggressions of Weakness

Ana Monteiro Ferreira, author of *The Demise of the Inhuman*, claims that micro-aggressions are the tools of the arrogant. In recent years we have not seen any politician with as many weapons of micro-aggression in his arsenal as Donald Trump.[8] He not only has those weapons but he is quick on the draw against any politician who might win glory from the masses. His dislike for the venerable John McCain and Mitt Romney may have come because of Trump's desire to be seen as the savior of the Republican Party.

Trump's adoption of the Ultra Right was his quickest path to political victory because the Ultra Right built populism on the basis of micro-aggressions against others. Trump saw the movement as needing a strong leader around whom all of the right could coalesce, Although Trump had lived a life that betrayed the values the evangelical Christians and white nationalists said they believed, he was able to influence them that he was, in effect, like them. Although Trump was a billionaire and many of them were mostly working poor and actual poor they felt that he could protect and support their interests. Demagogues always speak as if they are one with the people, that they have common interests, and that they rely on the people to give them direction. A demagogue therefore is not to be confused with a tyrant although there are similarities. Trump appears to want to act like a tyrant. He muses why he cannot order "his guys" to carry out his commands. He wants "his" people to do as he says even if it is not in the interest of the nation. He wants his cabinet minister to "serve" him because he identifies himself as the nation. Demagogues can morph into tyrants if they are allowed to do so by weakening democratic institutions.

Of Demagogues and Tyrants

Tyrants have a long history in Europe and throughout the world. Tyrants are rulers who seize power unconstitutionally or who are thrust into leadership by inheritance but then brutalize the people as if they are the only voice that could or should be heard. The best known of these types of rulers were Cypselus at Corinth and Orthagoras at Sicyon about 650 BCE. Of course, the tyrant would continue to appear in societies. So in Athens the title of tyrant was first given to Peisistratos, ironically a relative of the famous law-giver Solon. In 546 he succeeded to make himself tyrant after two unsuccessful attempts. Later, Lysander imposed thirty tyrants on the Athenian population after the capitulation to Sparta at the end of the Peloponnesian War in 404 BC. These men were dictators with no barriers and no bounds to do harm and to keep order by brutalizing the public.

There is always a danger when people give up their rational abilities to some "smart" leader. African Americans fought for the right to express our political preferences, name our own leaders, who would insure that our interests would be protected. If the interests are not even on the table they cannot be respected.

The Media and Democracy

The attack on the demagogue begins with relentless facts. As Michelle Obama said about Trump, he did not understand that America was a "rebuke of tyranny." But we know that the source of demagogic power is ignorance. One way to challenge demagogues is to shine the light of knowledge on their rhetoric. Once a people have elected a demagogue they cannot go back and ask him to read American history.

Those who would destroy democracy seek first to control the means of information. How do you inform the people? What do you know and where do you get the information? A cardinal part of the process of destruction of democracy is the rewriting of history. Thus, it is not only the present that must be distorted but also the past. The person who does not know the past will inevitably twist the present in order to shape the future.

Traveling outside the USA I normally do not listen to any news from the USA; I listen to the local news. Inside the USA one lives in a bubble of news that surrounds Trump. If it is on Fox it is pro-Trump, on most other networks it is critical of the demagoguery although it seeks to be balanced, which of course, is a false concept. Something is either true and correct or it is not. Can you imagine a world where Fox News is your only outlet to the world? You would be totally ignorant, misinformed, and hyperactive, believing that political reality is what you hear and see, often nothing but a maze of worthless conspiracy theories. This is the territory of the demagogue. It is as if someone had gathered all of the conspiracy theorists in one place and they repeat over and over again the same lines like parrots in a word cage.

A powerful and careful media is essential in a democracy. The people need to write commentaries, to inform each other of what is going on with elected leaders, and to refresh their own wells of information by reading, debating with respect, and writing with logic. The reason it was once said that democracy was based strictly on talk was because at one time people believed that whatever came out of your mouth should have been deeply reflected upon before you spoke. The Native Peoples believed that once you spoke you waited for a minute or two to allow your words to sink in so that if you needed to correct them before they were final, you could do so. There is a tendency among demagogues to believe that fast talking is a characteristic of intelligence when it is actually more likely the characteristic of lightweight thinking about substantive issues.

Every communication scholar and rhetorician know that a demagogue is a person who aspires to attain and maintain power by appealing to the base desires, prejudices, and fears of the public. A demagogue does not favor rational discourse and argument but rather seeks to control the actions and opinions of people by appealing to their worst instincts.

Finding out what people fear by opinion poll or survey or natural ability to listen to others who preach emotional rhetoric such as news commentators can give the demagogue ammunition for control. If you listen to commentators telling you how many foreigners are taking jobs away from your community, you may be inclined to listen to a demagogue who blames the foreigner rather than make a solid analysis based on economics. In fact, some people who have never encountered an international person will claim that they are unable to find jobs because of the international person. You do not have to see them to fear them becomes the maxim.

WHAT IS A DEMAGOGUE?

I believe that the demagogue is especially attuned to the psychology of the masses for his own benefit.

There are some classic definitions of demagogues and Luthin was quite succinct when he defined a demagogue this way: *"a politician skilled in oratory, flattery and invective; evasive in discussing vital issues; promising everything to everybody; appealing to the passions rather than the reason of the public; and arousing racial, religious, and class prejudices—a man whose lust for power without* recourse to principle leads him to seek to become a master of the masses. He has for centuries practiced his profession of 'man of the people.'"[9] Here it is, the person who knows flattery and is able to say with ease that someone is the greatest, most intelligent, finest, more gifted, and most beautiful, but never as great as the demagogue himself. Flattery is a coin of the demagogue. But so is invective a coin against those considered to be enemies or just plain opponents of the demagogue.

Historically, the word *demagogue* is derived from the Greek δημαγωγός which carries with it the idea of a leader of a mob, population, or organization. One of the great thinkers on this issue was Charles Wyatt Lomas, who taught at UCLA and wrote a book called *The Agitator in American Society* in 1968. Lomas studied a bunch of American political and social leaders and concluded that demagogues are most likely rabble-rousing speakers who can appeal to the prejudices of the masses in such a way that they identify their interests with those of the demagogue. However, he was clear that an agitator was not necessarily a demagogue but rather one who wanted to excite the people, especially those with grievances, to protest to the powerful to change a law, rule, or condition. The agitator, therefore, is an activist on the behalf of others.

On the other hand, the rhetorical power of a demagogue rests with the idea that all established protocols are against the people and must be overturned. The demagogue promises to do what others cannot do and would not do, thus the demagogue seeks to overturn protocols that he did not set up, to destroy ordinary customs of decent conduct, to disestablish the agencies, to corrupt the regulatory bodies, and to promise something entirely new and better. Rarely will a demagogue bring into existence anything better. If he promises better health care, you will get nothing better. If he promises that you will get tired of winning, you will see only losses. The point of the demagogue is to satisfy his own ego.

Using the most salient mark of the rhetorical process, persuasion, the demagogue understands that human beings have unlearned drives that can be manipulated toward narcissistic, nationalistic, and racist ends. Tapping into these drives can lead to certain negative behaviors. An effective orator or rhetor, in the event one may write, deals with rational discourse in an effort to persuade through argument. Hence, persuasion can be examined for its emotional, rational, and ethical content.

Demagoguery is coexistent with democracy itself, but it is never necessary for democracy. In many ways the political process that accompanies democracy is extremely vulnerable to the demagogue. So since the fundamental power of a democracy is supposed to be the masses of the people, not a court, not a legislature, and not even the rich oligarchs that might control the mighty technological empire of a nation, the demagogue can exploit the emotions of the people in order to gain access to democratic power. One might say that demagoguery hides in the corner of democracy. It is possible for people to give power to bullies who appeal to the lowest common denominator of a large part of society. One can never underestimate the desire of the masses to be identified with those they consider powerful. Trump knows that celebrity and wealth give him much more latitude to act as he does.

All demagogues know that the best way to appeal to people is to say that they are special, different in substance and nature than other people, and can be the salvation of a society in trouble if they take immediate action such as electing the demagogue to power. Most times the demagogue advocates an immediate and hardline agenda to show that he means business. In effect, the demagogue proposes something that a rational human being would question.

Enter Donald Trump the latest American to use demagoguery to win power!

Trump is a billionaire, but he convinced enough poor whites to vote for him because he understood their inner fears, their political desire to be exceptional, perhaps to feel superior to others. Consequently, Trump became a white nationalist leader as the president of the United States with almost no African American votes and only a few Latino votes.

There are two aspects to Trump's wealth, the foundation for his political rise; he is a *real estate mogul* and he was a *prominent television personality*. Combining money with celebrity Trump was able to rise above and to climb over all his political opponents.

Many people knew him from his real estate business because he built buildings and branded them with his name. These projects were hotels, golf courses, casinos, and office buildings. Trump branded and licensed his name to be used in managing this financial empire.

Trump also maximized his celebrity, when he became a political demagogue, by hosting *The Apprentice*, a television reality show, from 2003 to 2015. He also co-authored the book, *The Art of the Deal.* A businessman with a hard-nosed reputation for aggressive competition even if it meant the obliteration of the competitor, Trump managed to control two pageants, *Miss Universe* and *Miss USA*, for nearly twenty years. It is safe to say from comments made by some of his contestants that he saw the pageants as his personal sexual playground. Nothing he desired was ever off of the shelf and he felt that denial by a woman was not possible, particularly if he were a celebrity and wealthy. *Forbes* magazine called him the 544th richest person in the world with nearly 3.5 billion dollars, far less than some had thought,

Given the fact that he used his demagoguery to defeat sixteen Republican opponents to become the party's nominee for President, Trump's antics, and perhaps those of the Russians, increased his boldness in upping the demagoguery against Hillary Clinton. The official line has been that the Russians had no impact on the election. This is not credible because I personally know people who either did not vote or voted against Clinton after ingesting the bitterness spewed out from the Russians; and of course, Trump's willingness to amplify these negative messages showed his lack of restraint.

Trump had no hardline theories about politics except to appeal to the masses as an American protectionist, populist, and white nationalist. His followers immediately understood that he was someone who could ride with them toward a much more radically conservative country. Because of his wealth and celebrity, his populist views, often expressed with aflair, found immense free publicity and distribution. Running against Hillary Clinton, a formidable candidate with the best credentials of anyone who has run for the office, Trump used all of his political energy to defeat the progressive wing of American politics. Not only did Trump become the wealthiest and oldest person to assume the office, he became the most outspokenly narcissistic and nationalist president in a century.

The Appeal of Separation for the Demagogue

The African American community found him to be tone-deaf about the multicultural, pluralistic, and progressive pitch of the American nation. Of course, he had demonstrated his inability to accept African American leadership in his attack on Barack Obama. Trump's leadership of the Birther Movement sealed his toxic rhetoric and character in history and associated him with racial bigotry. He would also prove to have religious bigotry as well. Like a demagogue he would always claim that he was the least racist and bigoted person all the while demonstrating his dislike of Mexicans, Muslims, and people from Africa.

Collecting an unusual cargo of anti-Muslim, anti-Semitic, anti-African, homophobic patriarchs deeply rooted in the evangelical Christian faith, Trump went on his journey to steal the American democracy by employing the most vicious rhetoric seen in any recent presidential campaign. Neither Barry Goldwater nor George Wallace was ever accused of being so blatantly anti-democratic in their rhetoric. They were certainly not progressives, but they used symbolism, sentimental terminology, *double entendre*, and suggestion to persuade their audiences.

A demagogue seeks and gains power by creating a group of people who believe that he is a deliverer because he states that he is the only one who can make things happen. Demagogues play on the passions of the less educated and the poorer classes who are willing to give their trust to a self-promoting demagogue. He can claim that he is smarter, tougher, and wealthier than any others. No one can compare with the ego of the demagogue; he is supreme in his mind and seeks therefore to dominate all others.

Sycophants gravitate to demagogues in a waltz of mutual necessity. Demanding that those who "believe" in him give one hundred percent of their trust to him the demagogue is able to use and abuse the sycophant at will. On the other hand, the sycophant gains by being close to power. In some cases, as with Michael Cohen, the erstwhile lawyer for Trump, those close associates can try to capitalize on their access to the demagogue.

The American novelist, James Fenimore Cooper, as early as 1838, identified four basic characteristics of demagogues. Cooper claimed that demagogues had these fundamental features:

1. They fashion themselves as a man or woman of the common people, opposed to elites.

2. Their politics depends on a visceral connection with the people beyond ordinary political popularity.
3. They manipulate this connection, and the raging popularity it affords, for their own benefit and ambition.
4. They threaten or outright break established rules of conduct, institutions, and even the law.

Americans know James Fenimore Cooper as the first major American novelist but this New Jersey born writer was also a keen observer of history. Of course, the author of *The Pioneers* (1823), *The Last of the Mohicans* (1826), *The Prairie* (1827), *The Pathfinder* (1840), and *The Deerslayer*, (1841) was an astute student of human behavior.

However, most authorities believe that it was in his novel *The Bravo* (1831) where he explored the political corruption that he had seen and experienced when he traveled in Europe. Tyranny and political conspiracies created a patched fabric of bitterness that caused the masses to revolt against oligarchy. I believe that Cooper became a leading voice against demagoguery because he did not want his own nation to fall into the trap of endless struggles against demagogues and tyrants that he had witnessed during his travels in Europe. Of course, the United States was not immune from political intrigue as the presidency of Andrew Jackson, known as the 'Indian Fighter" also stirred the emotions of the masses. He was not, however, the model for Donald Trump although Trump placed Jackson's picture in the Oval Office. Not only was Andrew Jackson a racist slave-owner, he also oversaw the "Trail of Tears" that displaced thousands of Native Americans in order to give their lands to whites who made that land the basis of their wealth. Jackson sought a white republic for white people. Black people are not amused by Donald Trump's whistles to the most reactionary communities in the country. Demagogues dwell in the valley of difference; they do not see the possibilities in unity. One only has to see Obama's rhetoric as that of "no Red States and no Blue States, only the United States of America," to throw into relief the dumbing down found in Trump's jingoistic appeals to populism.

CHAPTER 2

The Fear of Fear

The demagogue trades in fear, but fear has little currency in the face of knowledge. We often think that this fear trade is what describes Trump's political rhetoric but obviously it has always been a part of his deal: scare the *other* person. In the summer of 1988 when he received an honorary doctorate from Lehigh University, Trump declared then that America was losing out to other nations because we did not have strong leaders. This show of a derisive and divisive personality trait seems to have been with him most of his adult life. His entire persona is to name those he likes and to dismiss those he dislikes especially stirring up the emotions of the lower classes, one of whom he is not, against the elites representing his own economic class. What he trades on is the fear that is in the hearts of many of the undereducated and uneducated that are left behind and economically depressed whites. There is the shock that they carry when they see the growth of immigrants, documented and undocumented; there is also a disdain that Africans, once enslaved in this land, are often able to cope with far less than whites. Trump understands these fears and he knows precisely how to play to them as a good populist. Whatever his political base populations are against; he is against. What they are for; he is also for until he changes his mind.

I surmise that the demagogue's tactic is to insure that he will have at least one supporter although no one is really secure standing close to him. Not only is agreeing with him or supporting him like "jello" as Senator Schumer once said about negotiating with Trump, but it is like soup. You cannot get in the pot with him without becoming dirty; that is the way Trump plots his life, seeking to overshadow all events and every other person.

But fear is not a philosophy or a political platform; it is a political tactic. The great warrior king of the Zulu, Shaka ka Sengakhona, is reputed to have said, "If you show your enemies and your friends your capacity to cause fear you will gain in immediate popularity." I sometimes think that this is the demagogue's greatest danger to any nation, that is, the ability to inflict punishment and assume that it will not return to you or your country.

Why did Trump ask the military to drop the Mother of All Bombs on Afghanistan when he took office? Dropping a 22,000-pound bomb, the MOAB, has been compared by the Pentagon to a small nuclear bomb. It is not even possible to drop it from a warplane it is so large; it has to be rolled out of the back of a cargo plane and outfitted with a parachute. Trump, I believe, wanted to show that he was willing to use the largest conventional bomb in America's arsenal to bring about psychological fear.[10] Of course, other demagogues are capable as well of demonstrating such wanton designs on human populations. So there is an immediate political bump perhaps in fear of the demagogue who is irrational enough to kill hundreds of people without remorse. Indeed, remorse is replaced by recrimination in the demagogue's playbook.

Such immediate response is no indication that a leader can control the longevity of fear. People can be fearful at one moment and not at another. Through study and analysis one can reduce the level of fear or understand what are the causes for fear. This emotion is nothing more than our anticipation that something or someone will cause us danger, harm, or hurt. Yet we can be in control of our emotions much more than those who preach violence against others or their followers simply because we understand our own actions. If I am peaceful or nonviolent in my actions I do not expect violence in return; yet I am certainly wise enough to know that it is possible that I may be attacked just as the peaceful Civil Rights marchers were beaten at the Edmund Winston Pettus Bridge over the Alabama River at Selma, Alabama on Bloody Sunday, March 7, 1965.

Looking at the plethora of cult leaders who have convinced people to follow them one quickly understands the power of fear. The true believer is a person who has abandoned rationality. This means that you have accepted the words of fear that comes from the demagogue without question. Trump appeals to three levels of fear in his followers: *economic, existential,* and *criminal.*

During his initial speech for his political campaign for the presidency Trump laid out what the public had to fear. On June 16, 2015 these are some of the ideas Trump stated in stark terms.

Economics

Lies.

> "When do we beat Mexico at the border? They're laughing at us, at our stupidity. And now they are beating us economically. They are not our friend, believe me. But they're killing us economically."

> "That's right. A lot of people up there can't get jobs. They can't get jobs, because there are no jobs, because China has our jobs and Mexico has our jobs. They all have jobs."

Criminality

> "When Mexico sends its people, they're not sending their best. They're not sending you. They're not sending you. They're sending people that have lots of problems, and they're bringing those problems with us. They're *bringing drugs. They're bringing crime. They're rapists.* And some, I assume, are good people."

Existential

> "Our enemies are getting stronger and stronger by the way, and we as a country are getting weaker. Even our nuclear arsenal doesn't work."

Trump's denouement was announced in these words:

> "Now, our country needs— our country needs a truly great leader, and we need a truly great leader now. We need a leader that wrote *The Art of the Deal*."

This is classic demagoguery of the first order. No one can do this job, he claims, but himself.

By way of contrast when Senator Barack Obama announced his candidacy for the presidency in 2007 he declared:

> "We all made this journey for a reason. It's humbling, but in my heart I know you didn't come here just for me, you came here because you believe in what this country can be. In the face of war, you believe there can be peace. In the face of despair, you believe there can be hope. In the face of a politics that's shut you out, that's told you to settle, that's divided us for too long, you believe we can be one people, reaching for what's possible, building that more perfect union."

Furthermore, his denouement closed out the announcement with this humble call for togetherness:

> "And if you will join me in this improbable quest, if you feel destiny calling, and see as I see, a future of endless possibility stretching before us; if you sense, as I sense, that the time is now to shake off our slumber, and slough off our fear, and make good on the debt we owe past and future generations, then I'm ready to take up the cause, and march with you, and work with you. Together, starting today, let us finish the work that needs to be done, and usher in a new birth of freedom on this Earth."

So the difference in speech is clear and one feels the inspiration, the hope, and the desire to lift all Americans in the deeply virtuous ideas presented in Obama's speech. Here, in Obama's tone and words there are no demons, no devils, and no enemies, only Americans with a myriad of challenges.

Democracy demands skepticism. A certain doubt or skeptical attitude will create a more viable democracy. The aim of some politicians in a democracy is to gain your vote, not to tell you the truth. Of course, we are fortunate that there has been a long tradition of politicians trying to satisfy their constituents without the use of fear. Others such as demagogues keep the people anxious, fearful of one another, and looking for enemies everywhere. At a moment when many Americans felt fear during the Second Great International War, President Franklin Roosevelt declared, "We have nothing to fear but fear it self!"

The demagogue loves fear because he is a bully. He enjoys seeing others tremble when they see him. He speaks to those who are weak as if he is the only savior for them. And weak people are always the ones who look toward the demagogue who declares how tough he is and how willing he is to speak on their behalf. We have watched as Trump used every device in the process of seducing the masses with his reality show celebrity.

Trump seems willing to sacrifice everyone and everything with his boasting about what should happen to his opponents. One does not see generosity in the demagogue; one only sees the outlines of a scared little boy who has better clothes, more money, and more materials things than his peers but he does not have soul, that ability to feel into the human condition in such a way as to demonstrate true empathy. Rather he uses the fears of his audiences for self-adulation and political advantage. There is something sinister in a person who shares almost nothing with those who follow him and he knows it yet he uses their explanations for their conditions as a way to dominate their thinking.

Trump is not a $29,000 a year man; he probably makes that amount of money in a few minutes if his reported wealth is what is claimed. Trump's followers are not his peers when it comes to membership in his clubs; it would take hundreds of them to find the money to pay the membership fees. Trump's health care is not the same as those who follow him. Demagogues cultivate reactionary followers who will fall for his lines against other people but they do not raise the level of reasoning, purpose, or American union. Therefore, the use of fear as a rhetorical tactic is venal.

In Trump's case it is not his wealth that is troubling but it is his attempt to claim that he is the champion of the poor whites that makes his rhetoric dangerous. It is false, fake, and artificial because it is an empty artifice used to manipulate the masses. He has shown an ability to use his mad-dog visage and countenance to trample on all the good other people achieve. He seeks adherents who will not question his judgment, tactics, or policies.

Trump did everything to criminalize those who did not support him during the election campaign of 2015. He attacked the Republicans who ran against him as weak, unintelligent, lazy, and low energy, thus making them different from himself, because he saw himself as intelligent and energetic. Trump viewed himself as the most perfect candidate to ever run for the presidency. This would be a theme he would continue during his first year in office as president.

Perhaps no losing political opponent had ever been assaulted as Hillary Clinton was during the campaign. During the time of the campaign Trump attacked Hillary Clinton with shouts of "Lock her up!" This was not enough; he also demeaned her character and spoke pejoratively about her health.

Trump had a particular animus it seemed toward African Americans. This may have been due to the discrimination charges filed decades ago by the National Urban League against Trump's properties. In February 2017 the FBI released nearly 400 pages of its investigation into Trump's discriminatory housing practices as far back as the 1970s.[11] Gerstein wrote, "In October 1973, the Civil Rights Division filed a lawsuit against Trump Management Company, Donald Trump and his father Fred Trump, alleging that African-Americans and Puerto Ricans were systematically excluded from apartments. The Trumps responded with a $100 million countersuit accusing the government of defamation." The case was eventually settled in 1975 with the stipulation that Trump institute "a series of safeguards to make sure apartments were rented without regard to race, color, religion, sex or national origin."[12] Of course, by now Trump had established a pattern related to African Americans. When on April 19, 1989 a young white woman, Trisha Melli, was assaulted, raped, and sodomized, the tragedy infuriated the city. The investment banker had been jogging in Central Park when she was attacked and then left in a coma for 12 days. Four black youth had been apprehended on the night of the attack. The investigators discovered that none of the boys' DNA matched with what had been discovered on the victim. Yet the public outcry led by politicians and noticeably by real estate builder Donald Trump was for the blood of the young black men. Within days Donald Trump took out a full-page ad costing nearly $100,000 in four newspapers calling for the return of the death penalty. He said, "Muggers and murderers should be forced to suffer and, when they kill, they should be executed for their crimes." They were convicted and given sentences ranging from 5 to 15 years in prison.

Yet even when in 2002 a serial rapist whose DNA matched that found on the victim was arrested and confessed to the crime, Trump refused to back away from his earlier vile claim. In fact, he doubled down after Matias Reyes was convicted of the crime saying the four young black men were guilty because they told the police they did it. Investigators found that the young men had been coerced into confessing a crime they did not commit. Trump refused to apologize saying later, "They admitted they were guilty," he told CNN. Still not convinced of the innocence of the =young men Trump continued, "The police doing the original investigation say they were guilty. The fact that that case was settled with so much evidence against them is outrageous." What was outrageous is Trump's irrational response to the innocence of these young black men. It is not as if he had not ever heard of false confessions where a disproportionate number of black men are coerced into confessing crimes they did not commit by the police. In a more recent example of his animus Trump told National Football League owners that they had to come down harder on the players who expressed their dismay that the nation did not rise-up against police brutality. In other words, Trump had no sympathy for the thousands of young black men who are brutalized every day; he acted if he did not know that the protest was not against the flag but against police brutality. Colin Kaepernick had started the protest as a nonviolent response to the litany of murders of black men on the streets of America by the police. Trump's reaction was to weaponize patriotism and to suggest that the young men were doing something un-American when what they did by kneeling was ultra-American. Does Trump side with the victims of brutality and does he show empathy toward the players? No, he rants "*Wouldn't you love to see one of these NFL owners, when somebody disrespects our flag, to say, "Get that son of a bitch off the field right now. Out. He's fired. He's fired!" You know, some owner is going to do that. He's going to say, "That guy that disrespects our flag, he's fired." And that own13er, they don't know it. They don't know it. They'll be the most popular person, for a week. They'll be the most popular person in this country.*"[13]

A demagogue finds pleasure in recriminations and spurious arguments; it works him up and allows him to express himself in the angriest of tones. Catharsis is the idea that people can find relief from repressed emotions if they discover someone who can show them that it is alright to hate, to enjoy suffering of others, and to do harm. Trump knows how to use negative images and outlandish ideas to evoke these repressions.

Genetic Superiority

Seeking comfort in the belief that one is smarter, more intelligent, or more capable than others is a luxury demagogues love to own, whether it is false or true. Trump starts with the classic identification mark of the traditional racist: claiming that his blood is "good stuff." Trump boasts of his German blood not realizing that there is no such entity. In fact, the German people are not different from other Europeans since according to scientists the ancient migration routes have been mixed for thousands of years. We know that the vast majority of Europeans, including Germans, descended from three major migrations during the past 15,000 years. Originally all *homo sapiens* can be traced back to the Mother Continent, Africa. Migrants have a history of mingling with other people in a sort of remix of cultures.

So where does Trump's pride in "German blood" come from if not from his identification with German culture? Even if he knew German history he would now know that Germany has expressed itself over the past sixty or more years as a nation of *wilkommenkultur*. Trump is not interested in this progressive and inclusive Germany since he wants something that makes him different from other human beings.

Trump does not just talk blood; he talks also genetics, claiming that he has genetic superiority. In a moment of unscientific insanity he said, "All men are created equal – that's not true. When you connect two racehorses, you usually end up with a fast horse. Secretariat doesn't produce slow horses. I have a certain gene. I'm a gene believer. Do you believe in the gene thing? I mean I do. I have great genes and all that stuff, which I'm a believer in."[14]

Trump believes in the racehorse theory. Even more, he believes that some people cannot produce "fast" offspring. He is a racist at his core. This is why he really fears the influx of people from non-Northern European nations. The browning of the American population has been going on since the turn of the last century and Trump sees himself as the doorstopper for unfettered immigration. This is the same old odd understanding of nation as other racists.

Nations are not static; the United States is not today what it was one hundred years ago and it is not today what it will be a hundred years from now. I do not fear that one day the nation will be primarily Spanish or Chinese speaking. What a patriot wishes for is the maintenance of good values, common sense, self-affirming, and dignity producing ideals. In the American context a nation is not defined as "blood and soil" but by unifying experiences and collective purpose.

One of the most terrifying attributes of Donald Trump is that he demagogues anything and every idea placing himself at the center of all discourses. As we have seen in terms of intellectual ability he thinks more highly of himself than he ought. Consequently his language is reprehensible for its self-centeredness and egotism. Anyone who talks excessively about his own importance because of an undue sense of self-regard might be considered an egomaniac.

Trump says, "Well I think I was born with a drive for success. I was born with a certain intellect. The fact is you have to be born and be blessed with something up there. God help me by giving me a certain brain. It's *this*, it's not my salesmanship. *This* – you know what that is? I have an Ivy League education." He spent his last undergraduate year at Wharton, the business school of the University of Pennsylvania. "I have like a very high aptitude." Trump tapped his head, as an adolescent might do to indicate intelligence, while saying all of this on CNN.

Trump announces his beliefs, much as all demagogues do, because he knows that there are certain people who will share these beliefs with him. Some will defend his comments as true even though they do not like the fact that the demagogue expresses himself so clearly. They give him a pass because he is bold enough to express their prejudices. Thus, following his racehorse theory Trump says over and over again something like this:

"You're born a fighter, and I've seen a lot of people who want to fight but they can't. Some people cannot genetically handle pressure. I always said that winning is somewhat, maybe, innate. Maybe it's just something you have; you have the winning gene. Frankly it would be wonderful if you could develop it, but I'm not so sure you can. You know I'm proud to have that German blood, there's no question about it. Great stuff."[15]

The demagogue is a frightful person. He operates from a sense of inferiority and the expression is bullying because he wants to convince himself that he is superior. This is fear, pure and simple.

Trump says "Winning is innate." His many followers, some who are undereducated believers in white supremacy, come to the same conclusion as Trump. They believe that to win you have to have a superior genetic gift, something given to whites at a higher distribution rate than any other human group.

Trump believes that he was born with special bra
him good at English, at handling pressure, and at wi
there are hardline votarists who will not abandon
to Fifth Avenue in New York and shot someone, as he shock
political rally, is the essence of fear.

Most demagogues create their own reality. Actually, if they lived in an authentic environment where they were conscious of others they would see that all around them were examples of "really smart" people who do not have illusions and who come from every ethnic group. The attack on Mexicans, Africans, Arabs, and anyone who is not Nordic or Aryan means that Trump is a racist. Why would he not want for others what he wanted for whites?

Only the Native Americans are not children of the ships. This is why it makes no sense for the politicians to try to demagogue immigration. That is, if you understand the origin of the nation as the United States of America you know that it has nothing to do with "blood and soil" since the overwhelming majority of the people who call themselves Americans came from other continents. But there are two elements in American society that have been exposed in Trump's America. The first is the fear of strangers and the second is the doctrine of superior whiteness. of course, the first strangers to America were whites who landed on the shores of America and were greeted overall with hospitality as far as we know until the Europeans sought to impose their values, culture, and ideas on the land of the Native Americans. This was the beginning of conflict. Jack Weatherford wrote a book called *Indian Givers: How the Indians of the Americas Transformed the World*, in which he argued that most Americans know next to nothing about the Native peoples.[16]

In fact, Weatherford argued that the concepts of egalitarian democracy and liberty were already in the Native American's vocabulary for governance before the encounter with Europe. These concepts are not derived from Greco-Roman or French derived culture; they came into being because of Native American ideology and were interpreted and translated into European language and culture.[17]

are excepted because they are acceptable. They have the right blood and there is no *rassenschande*, racial crime, in mixing with Norwegians.

What must the demagogue convince the people to fear? In Hitler's case it was the Jewish people, the intellectuals, and the media that they did not control. They used every controlled media, especially filmmakers at the time, to propagate the ideas that the male Jew was a sexual predator, Jews were like rats, and the way Jews killed animals for kosher was distasteful. Films such as *The Eternal Jew* was meant to remind Germans that they had to work for a society of pure German blood.

As it was in Germany so it was to be in Trump's rise to power in the United States, the drumbeat for an American exceptionalism that carried with it racist and religious overtones would become the favorite coin of the demagogue.

White nationalists, racist groups, and right-wing protesters converged on Charlottesville, Virginia, on August 11-12, 2017 using a Nazi-rallying cry. Video showed that some of the protesters shouted "blood and soil," a phrase invoking the Nazi philosophy of "*Blut und Boden*." This was the Nazi ideology that stressed ethnic identity and nationalism based on blood descent and the territory. The German farmers and peasants celebrated this idea as a virtue for their nationalism.

Of course, in the United States of America the only people who are considered indigenous to this land are the Native Americans, not the immigrants from Africa, Asia, or Europe.

Regardless of the historical facts, the group that gathered to protest Charlottesville's plan to remove relics of its Confederate past, such as a statue of Confederate Gen. Robert E. Lee, was bent on a torchlight march. They clashed with counter-protesters who promoted calls for "respect, freedom, and unity." The counter-protesters also swamped the reactionaries with a multi-cultural and multiracial progressive group of marchers who confronted the Nazis and white nationalists with a call for national unity without symbols of hatred or false ideologies glorifying slavery. It portends the possibility of a society that will overcome fear and the demagogue.

Chapter 3

The Democratic Ideal

Representative Adam Schiff, California Democrat, ranking member of the of the House Intelligence Committee, said at the University of Pennsylvania, on February 1, 2018, that President Donald Trump was "not a champion of democracy."

We all learned in high school or college that Athenian democracy appeared in the fifth century BCE in the Greek city-state of Athens.[19] It extended to the area surrounding the city called Attica. This was the first identified democracy in the Western world. It has a long and idealized history.

But demagogues have always challenged democracy and the system has demonstrated its resilience although it has not gone without exposing its possible fractures caused by those who seek to destroy its effectiveness.

The longest serving democratic leader in Athenian democracy was Pericles who made the Delian League an empire and later led that Athenian empire during the first two years of the Peloponnesian War. Pericles was a popular leader who led the citizens in the arts and literature and made Athens a center of learning.[20] Considered one of the great orators in Greek history, Pericles Funeral Oration celebrating fallen soldiers is one of the most remarkable speeches ever given. Listen to the brilliance and eloquence of the democratic Pericles:

> "Freedom is the sure possession of those alone who have the courage to defend it."
>
> "Just because you do not take an interest in politics doesn't mean politics won't take an interest in you."
>
> "What you leave behind is not what is engraved in stone monuments, but what is woven into the lives of others."

Trump is the anti-Pericles. He has no ambitions to make the people of the United States more intelligent and more democratic. There is not an element of eloquence in Trump's heart. True eloquence is good speech combined with passion and truth. Trump is no Pericles. His highest ambition seems to be to secure "love" for his brand of a person, but ultimately this is a losing battle because that which lasts is indeed woven in people's hearts.

The Age of Pericles is the opposite of the Age of Trump. When only a third of the American people approved of Trump doing his first year as president he claimed far more than he had achieved and apparently plotted to retain his office through a personal universe where he thought the people were stupid.

As bizarre as it may seem to some, most Americans *know* stupid and they know deplorable; democracies react to demagogues who seek compliance at all levels. No demagogue has ever secured a lasting positive reputation in history. The poisoned self-interest of demagogues is the seed of their destruction. Trump's political end and moral demise represent a flaw in his character, not a failure of the system.

The biographer Plutarch tells the world that Pericles stood first among the Athenians for forty years as leader of Athens from the time he was about 35 years old. The middle of the fifth century belonged to the Leader of the Democracy. Pericles was not flamboyant. He became a great model of judicious decisions, wisdom, frugality, and privacy, evidencing the ultimate judgment of history on those leaders who try their best to promote decency, justice, harmony, unity, balance, and order.[21]

Inexact and fractured as it was from the beginning, the Declaration of Independence, written principally by Thomas Jefferson, held within it the promises that could become the legitimate bases for protest and progress. On the other hand, the Constitution of the United States fathered by James Madison outlined the form of government for the people of the United States.

Thomas Jefferson and James Madison envisioned a system of government of the people, by the people and for the people where eligible people are elected to represent the people. In effect, they were flawed democrats and capitalists whose philosophy constantly conflicted with their actions.

We are able to feel a sense of personal growth in the intellectual and moral character of Jefferson whose most famous quotes may be these:

> "The tree of liberty must be refreshed from time to time with the blood of patriots and tyrants."
>
> "Honesty is the first chapter in the book of wisdom."
>
> "In matters of style, swim with the current; in matters of principle stand like a rock."

If you only examined Trump in the context of these statements you will see the difference between a demagogue and democrat. The tree of liberty must always be protected from demagogues by the vigorous defense afforded by rational thinking. Furthermore, lying to the public as Trump has done thousands of times will eventually unsettle the masses to the point where he will be chastised by the democratic process. One thing seems certain and that is that the democratic norms, though buffeted by malicious turbulence, are resilient.

Patriots must have some skin in the game, and during the rise of Trump we have seen very few patriots on the side of the Republicans willing to fight a demagogue; they have been most willing to acquiesce. Yet a few of the Republican senators, especially Jeff Flake and John McCain both of Arizona, have warned the nation of the political climate created by Trump's antics and ideas.

In October 2017 when Arizona Senator John McCain received the Liberty Medal for public service from the National Constitution Center in Philadelphia he warned against the United States moving toward "half-baked, spurious nationalism." From his political platform as one who has been in national leadership, McCain warned that America should not fear the world that the USA has led for a nearly one hundred years by abandoning "the ideals we have advanced around the globe." McCain said that the United States should not "refuse the obligations of international leadership ... for the sake of some half-baked, spurious nationalism cooked up by people who would rather find scapegoats than solve problems."

Trump's "State of the Union speech" was generously sprinkled with elements of narcissism and authoritarianism. As a demagogue he tries to place himself in the position of determining what others should do and if he senses that people are in opposition then he attacks like a wounded animal.

Trump called Democratic legislators treasonous simply because they would not applaud a line in a speech where he claims something that was not originated on his own watch is to show the simple gall of the demagogue. Why should those who know that he did not create the conditions for African American employment to improve stand and applaud his usurpation of Barack Obama's legacy? The willful lying and creative obfuscation of the truth show the dangerous capabilities of a person whose values all reside in self-praise.

So Trump gives his first State of the Union Address, and could not resist the urge, to wrap himself in the fabrics designed and created by another president? In many ways it reminds me of the fascination that Turkish president Recep Tayip Erdogan has regarding Ataturk. When I visited Turkey traveling from Istanbul to Ankara and back to Bodrum everywhere I went I felt that the great Ataturk was under threat by the current president. The fact is that Ataturk, the great reformer of Turkey, had been dead since 1938! Trump's attack on his predecessors was almost immediate. Trump's speech to the USA Congress was filled with shameful rhetoric by a would-be demagogue who felt threatened by opponents who did not give applause to his proposals.

Senator Jeff Flake (R-Ariz) condemned President Trump for saying that Democrats who didn't applaud his State of the Union address were "treasonous." Flake added, "Treason is not a punch-line." As a demagogue would do Trump took the lack of applause for his speech by Democrats as a personal insult; he said that they were "un-American."[22]

Senator Flake also claimed that the demagogue did not know the meaning of the word "treason" and furthermore did not appreciate the power of presidential words.

"I have seen the president's most ardent defenders use the now-weary argument that the president's comments were meant as a joke, just sarcasm, only tongue in cheek," Flake surmised.[23] Then he continued, "As members of Congress, we must not ever accept undignified discourse as normal because of the requirements of tribal party politics."

The tone for negative discourse, dishonesty, narcissistic rhetoric is set by Donald Trump in his demagogic manner. Lying remains at the core of all Trump's rhetorical misdeeds. He is skillful at creating false information, disinformation, and half-truths in a way that has rarely been seen in presidential politics.

If honesty is the first chapter in the book of wisdom, then Trump is wandering around outside of the library. He has not opened the introductory book of wisdom because he has become synonymous with liar. Trump apparently sees lying as a functional part of governing because with the lie one is able to follow the pessimistic strategies of the late Roger Ailes and Roger Stone. They became infamous for creating political drama that could be followed by political audiences. Roger Ailes had been a consulted for Richard Nixon, George H. W. Bush, and other Republicans, including Rudy Giulliana. He is credited as one of the initiators of the Orchestra Pit Theory that offers this question, "If you have two guys on a stage and one guy says, "I have a solution to the Middle East problem," and the other guy falls in the orchestra pit, who do you think is going to be on the evening news?"[24]

Watching for Despots

We are accustomed to calling James Madison the other brain of the founding American documents. The *Declaration of Independence* was written in 1776 and the Americans subsequently created the *Articles of Confederation* in 1777 to hold the country together. This Confederation would last for ten years and then the Constitution of the United States of America would be written in 1788 and ratified in 1789. The aged Benjamin Franklin was asked at the close of the Constitutional Convention as he left Independence Hall on the final day of deliberation in 1787. "Well doctor, what have we got—a republic or a monarchy?" Franklin stared squarely in the person's eyes and said, "A republic, if you can keep it."

Trump is neither a Jefferson nor a James Madison. Jefferson was an intellectual, gifted with the ability to marshal his talents to construct an eloquent response to the British Empire, despite his own collusion with the enslavement of Africans and Madison found the strength of character and the mental agility to write *The Federalist Papers*, alongside Alexander Hamilton, despite his inability to rein in Andrew Jackson's racist proclivities against the Native Peoples, but Trump has defined no intellectual theory of governing, no political ideology, and no philosophical curiosity about how to make a nation better. He has rather acted like a despot even deliberately misleading the public with calls, for example, for the investigation of the FBI and CIA. Trump demanded that he wanted the Department of Justice to investigate if a spy was embedded in his campaign. In fact, he "demanded" such an action in the same way that some Eastern European autocrat would direct his underlings to act. Trump sees government as one man's complete rule, but democracy is by definition the rule by the people.

Demagogues are not known to be thinkers; they are merely narcissistic actors. On the other hand, dictators have an interest in governing and in the theory of government; they could be beneficent or malevolent.

Because Trump has no particular insight into political theory or history he can easily become an unwitting puppet of either foreign operatives or the American oligarchs who represent his class interests. I am not even certain that he cares about the writing of laws since he seemed so distant from every piece of legislation that landed on his desk. He is simply a rubber stamp in search for numbers to feed his hunger for political love. Perhaps the one action that he has taken as a consistent task is to reverse every Obama Executive Order and to assert his own orders.

Power attracts many people that are easily beguiled by the quest, and still others are corrupted by holding office, and many of those are not worthy of the power that they gain. Misuse of the public's trust is almost predictable because oligarchs, especially those who are wealthy, waste no time in drinking from the wells of the public to make them richer.

I recall how elegantly and eloquently former President Barack Obama pleaded with Americans to defend democracy in his farewell speech in Chicago on January 10, 2017. In his own words Obama said: "After eight years as your president, I still believe that. And it's not just my belief. It's the beating heart of our American idea – our bold experiment in self-government. It's the conviction that we are all created equal, endowed by our creator with certain unalienable rights, among them life, liberty, and the pursuit of happiness. It's the insistence that these rights, while self-evident, have never been self-executing; that we, the people, through the instrument of our democracy, can form a more perfect union.

This is the great gift our Founders gave us. The freedom to chase our individual dreams through our sweat, toil, and imagination – and the imperative to strive together as well, to achieve a greater good."

A demagogue is incapable of such words because a demagogue does not believe in these fundamental ideas. Rather than strive for the greater good the demagogue strives for his own glory pushed along by the "dilly-dilly" of his acolytes. The country's first declared black president was elected in 2008 on a message of hope and change; he was then re-elected in 2012 as the society showed an authentic desire for a democratic state. He had ambition, confidence, and boldness in the face of issues, but he was neither petty nor unwise.

Donald Trump, vowed to undo many of Mr. Obama's signature policies but as former Vice President Joe Biden said on February 6, 2018 on an interview with Chris Cuomo, Trump will not be able to reverse all of the achievements of the Obama years but would try to do his best to damage the goods.

One of the more controversial proposals Trump made during the first part of 2018 was for a military parade like the ones in France, China, Russia, and North Korea. Not only is this something that runs contrary to American tradition over the past fifty years; it is something that Trump embraces from the most authoritarian part of his brain. He is a demagogue who seeks military exhibitionism as a reflection of his strength. Showing off military might, regardless to how much money it will cost is merely a display of his bravado.

Patriarchy and Arrogance

Trump has a patriarchal temperament that is expressed in his arrogance in attempting to discredit the Department of Justice and the Federal Bureau of Investigation. But it is also a part of his pomposity derived from his running of beauty pageants and working as a television personality. Nothing escapes his patriarchal tenacity. Trump does not worry about anything but nationalism and populism and hence he tries to divide the people, us against them, one against another, in order to maintain his control over a declining political base.

A demagogue is not a role model and must never be seen as someone whose behavior is admirable in spite of Trump's declarations that people voted for him even when they had heard about his misdeeds and attitude toward women. A demagogue gathers all moral judgments from himself to himself; he is wrapped in a robe of incredulity. One hears him, sees him, and rushes to check with neighbors and friends and family that you have heard precisely. What gives Trump, the demagogue, the authority to say that when people "deny" that they sexually harassed women they must be given a pass? It is nothing short of a patriarchal prerogative in Trump's mind. He has no moral compass as one would expect of a president of the United States.

Patriarchy has distorted the nature of right and wrong in the political era of Trump. In fact, Trump sides with those who act like him, talk like him, and think like him. A demagogue learns quickly that there is a certain segment of a democracy that will acquiesce to any type of leadership. There is nothing in history that tells us that a demagogue will not be acceptable to a certain minority of the electorate. They see in the demagogue the "Father Figure" who will take care of every problem and all issues that have plagued society. It is the glow of the patriarch that covers the inadequacies of the votarists.

All indications from those who have accused Trump of seeking to degrade women are that he is a regular or ordinary abuser of women, that means he does not believe the stories of women victims who accuse men of harassment and domestic violence. He defended Bill O'Reilly, Roger Ailes, and Roy Moore and other alleged abusers by saying that they "denied" the accusations. For him, the idea is to "deny" and you will be likely to beat any charges, at least, in the public eyes. One could say, "he denied it." Of course, several women have accused Trump of sexual harassment. These violations remain one of the complications for the Trump presidency, particularly if the Democrats win elections in 2018 and retake the Congress. Trump sees men who are patriarch as allied with his views of society.

Patriarchy is a system of power, men over women, that contains the elements of hegemony and domination. As a political system patriarchy excludes women or keeps them in minor roles because men see themselves as the natural leaders of government, society, and the family. With such a male focus the demagogue, who celebrates the domination of women, resonates with those who see exploitation of those considered weaker as natural. Such a position is the essence of gender chauvinism.

Great nations throw off despotic governments. We are always responsible for our political condition and when we are confronted with despots we must refuse domination. Trump aims to muddy the waters of national discourse by lying about anything that will make him look bad. He apparently does not care that lies are the source of much of his political problem;

however, the reaction to Trump has been strong from those who believe that he is a danger to democracy.

Chapter 4

The Fragility of Civility

Americans have come to accept the President as the Commander-in-Chief as well as the moral leader of the nation. Sadly, Donald Trump has shown himself to be out of his league as both a respected commander and as a model of morality. When you cannot depend upon the president's words, or take cues about dignity, balance, order, race relations, and values from the president actions who speaks and behaves in public as a hooligan and a bully, it demeans the public and creates a society that is rough, rude, and disrespectful in relationships. My postman said, "If the president is an idiot then the people who elected him must be idiots." I would never go that far but I do believe that If the president is a liar, then those who elected him, knowing that he lies, must also accept lying as a matter of principle. If the president is a misogynist and homophobe, then those who elected him must harbor those sentiments, we believe. So the fragility of the American civil society is in the hands of diverse institutions but the institution of the presidency is certainly one of the most important when it comes to sustaining principled government. The slow slide into corrupt, immoral, petty political machinations has been going on for a while but Trumpism has accelerated the demise of decency.

Let's examine how Donald Trump relates to others. Here is a partial list of negative nicknames and epithets he has used:

- Dicky Durbin for Senator Richard Durbin, (D-Ilinois)
- Al Frankenstein for Senator Al Franken, (D-Minnesota)
- Sneaky Diane Feinstein for Senator Diane Feinstein, (D-California)
- Sloppy Steve Bannon for Trump's own Chief Political Strategist
- Low Energy Jeb for former Florida Governor, Jeb Bush
- Little Marco for Senator Marco Rubio, (R-Florida)
- Crying Chuck Schumer for Senator Chuck Schumer, (D-New York)
- Sleepy Eyes Chuck Todd, NBC and MSNBC, Meet the Press host
- Lyin' Ted for Senator Ted Cruz, (R-Texas)
- Crazy Bernie for Senator Bernie Sanders, (I-Vermont)
- Crooked Hillary for Deimocratic nominee, Hillary Clinton
- Psycho Morning Joe for former Republican Congressman Joe Scarborough
- Little Rocket Man for North Korean leader, Kim Jong Un
- Pocahontas for Senator Elizabeth Warren (D-Massachusetts)
- Jeff Flakey for Senator Jeff Flake (R-Arizona)
- Liddle Bob Corker for Senator Bob Corker (R-Tennessee)
- Wacky Congresswoman Wilson for Congresswoman Fredericka Wilson
- Little Adam Schiff for Congressman Adam Schiff (D-California)

There are scores of other names and epithets used by Trump to vilify his opponents and to entertain his followers, but when the hammer of public opinion or legal judgment falls on him he will find few sympathizers among decent people. He has provided his followers with an account filled with unsavory characteristics and embarrassing personal conflicts. Trump has continued to cultivate a loyal patronage, to erode institutional integrity, to direct attacks on security organizations, to debase civil servants, to ridicule those who trust the political system, and to undermine democracy by not confronting the Russian interference in American politics since his election. Indeed, his entire behavior as a campaigner and a president fits the pattern of a very petty person who could never consider any other person equal or better than himself. Yet as Bill Gates, one of the richest people in the world, and one who is highly respected, pointed out, Trump did not know the difference between HIV and HPV. Trump's pettiness allows neither the ability to learn nor the ability to respect others.

This is one reason why Trump never gave a positive nickname to any other president. Clinton, Bush, and Obama were all bad in his opinion; only he was a superior person, only he knew how to be president. Everything that he could think of about opponents seemed to be negative, and his language was filled with dignity-demeaning terminology. Running as fast as he could away from dignity-affirming categorization Trump lands on the edges of negativity.

Legitimizing a demagogue is an insult to an informed democratic polity. Yet the demagogue uses the cultural connection to the conservative movement to articulate his views. The vulnerable nature of democracy is revealed in how easy a foreign power using the pathways establish by the people themselves can invade the electoral process. A demagogue, scared that admission of a foreign government's interference, will question his election, claims that it is a witch hunt for the department of justice to examine the Russian role in American elections although more than 70 people and organizations have been charged as of 2018.

With an objective to knock out the pillars of democracy and the government institutions that place constraints on the executive Trump started his administration stating that he did not believe all of the reports of Russian spying and involvement in the American elections.[25] He demonstrated disrespect for all of the CIA operatives who had died on the job.[26]

Trump without any analytical documentation or critical appraisal claimed that he wanted to reconstruct the FBI and the Department of Justice. One is not clear if Trump is simply setting up actions that he could claim as something done by his narcissistic self or if he actually believes that these hallowed institutions need change. Governance is always best when there is deliberation and this means consultation with others who are experts. It does not mean that an executive could not act; it simply means that the executive should act after advice.

In some senses the electoral process, a coin of modern democracy, is also one of its weakest links when it comes to the ambition of the demagogue. It is possible that democracies will be attacked in the future by a conspiratorial strategy of demagogic rhetoric and technological intrusions. The adaptation of assaults on democratic institutions will demonstrate the vulnerability that is built into the American system. Fierce battles to control the mechanisms of electoral politics will catapult the demagogue into a clearer view where the profile of ignorance, catastrophic divisiveness, and crudeness will render the society *awoke* to social and political trouble.

History is no joke; it always have lessons to teach. Adolf Hitler and the National Socialists lost the presidential election of 1932 in Germany but President Paul von Hindenburg later appointed Hitler the Chancellor. He moved quickly to change the cabinet and to clamp down on opponents. Hindenburg later paved the way to dictatorship and war by issuing the Reichstag Fire Decree that nullified civil liberties. After a series of parliamentary elections and associated backroom intrigues Hitler emerged as the supreme leader surrounding by a cadre of his political operatives willing to struggle to attain and maintain power in Germany. Raiding the various ministries and the cabinet of politicians and civil servants who had served previous governments, Hitler was able to root out many anti-Nazi personnel. Trump's notion of a "deep state" may have come from some of his nationalistic strategists who have seen the progressive tendencies of a liberal leaning state structure operating in the interests of civil rights, health rights, protection of nature, and anti-corporatist policies that crowd the rich with more wealth while failing to support the good of the public order by taxing the wealthy to build a better society. Getting to the bottom of this "deep state" has meant loading the judges' seats with some who are white nationalists, anti-people politicians, and racists who neither love nor respect the diverse society that the United States has become.

The demagogue enjoys using militaristic and physical metaphors and examples. When Trump as candidate for the presidency spoke of beating up, knocking out, pushing arrested people into police cars without protecting their heads, and calls to "lock her up" in reference to Hillary Clinton he was exhibiting one of the features of the authoritarian personality. Trump strikes either when he feels under siege or when he wants to knock out his opponents before he is under siege. In other words, his entre character is wrapped up in the martial posture of attack. He is supported or he supports himself by dragging out fake news stories and then condemning the mainstream news companies as the purveyors of fake news. It is a sinister political and rhetorical assault on the foundation of what is considered the basic principles of a democratic society. For Trump, all news should show him as he sees himself, that is, as a great person or else it is worthy of attack.

Civility is a thin thread in human society; it is not a rope that may hold even when it has to bear the weight of thousands of lies. Civility breaks in the face of constant propaganda of lies. It causes people to want to fight, to defend their dignity, and to recoil at the slightest provocation. The use of fake news by a demagogue confirms that he has no sympathy or empathy for those who are mangled in the iron jaws of lies. A demagogue will sacrifice the institutions of liberty for the sake of his own self aggrandizement. All conversation and actions are important to the demagogue only so far as they underscore his special genius.

Donald Trump mastered the propaganda technique of getting ahead of his opponent by announcing that any news that showed him in a negative light was "fake news." Of course, he also criticized information that demonstrated his policy ideas were not grounded in reality. The problem with demagogues is that all news that aims to set the record straight after they have lied is called fake news. But what is fake news?

Fake news is a specie of propaganda that uses misinformation, lies, fabrications, and hoaxes to blind people to the truth. A fake news operator is likely to throw so much misinformation into the atmosphere that the person receiving the news will be completely befuddled. A brand of fake news is broadcast, printed or distributed over social media as a deliberate instrument of political confusion. Gaining politically or financially from falsehoods spread abroad is the main purpose of fake news. Those who intentionally mislead others with information that is inauthentic and used only to serve the political interests of the demagogue are the worst attackers on the democratic institutions because they corrupt the very ground upon which the trees of security and respect grow in society.

A particularly dangerous aspect of the "fake news" controversy since the days of Trump has been the fact that the demagogue has mastered the technique of creating his own reality and then passing it off as if it is real and all other information, however gathered by reputable media agencies, is fake. One can only do this with a certain type of boldness in order to convince the masses that what they see they actually do not see. It is like his defense of the sexual abusers who hide behind denial.

The demagogue succeeds at this because all demagogues master human behavior. During his rise to celebrity Donald Trump, it is claimed, often called in to radio stations or other media to give the host positive reports about himself. He knew the habits of media institutions and the need for something new and exciting all the time. Thus, he would disguise his voice and call a station supporting ideas that Donald Trump was smart, great, and intelligent wen it came to business. For people who did not know that Donald Trump was a manipulator of the media they found him to be credible and a person of good will.

Demagogues also know that ratings bring in advertisers and that advertisers make money for the media companies. We now know that Facebook News Feeds and other media instruments have played into the hands of the fake news creators. There are several dangers with this scenario, each one capable of taking down a government or a president. The first is the danger of the ease with which young people can get access to false information often contradicting what they learn legitimately in school. The second is the cooptation of the news feeds by hostile actors on the world stage who create false information and market it by expressing it on legitimate news services as if it is real. Almost no American demagogue before Trump has had such easy access to millions of people. His calculus has to be that his millions of followers on Twitter who are exposed to his lies and created stories will never abandon his leader.

I have found that fake news now competes with real news for the attention of large groups. During the 2016 presidential campaign fake news groups probably from a hostile nation sent out false political information to African Americans in an attempt to disrupt the strong Hillary Clinton advocacy group. It was reported in May 2018 that Steve Bannon, erstwhile Trump advisor and white nationalist, had worked to suppress the black vote in the 2016 election. Although we know that many people acted on this information whether to hold rallies or to spread the anti-Clinton propaganda through social media we are yet unsure about whom to precisely implicate. Russia has been accused but it has been suggested in some circles that the Russian trolls may have operated with the support explicit or implicit of the Trump campaign. I believe that Buzzfeed found that many fake news stories about the presidential campaign of 2016 had received more social media interaction than the top 20 legitimate news stories by reputable companies. One explanation of this credibility and information phenomenon may be that the fake news items are not fact-checked, edited, or discussed in editorial boards. In other words, there are no rules to which the fake news hawker has to adhere. In these cases the American masses are like ducks in a pond being fed anything that is thrown in by the passers-by. How do you tell bread from crud if it has the same color and consistency?

Fake news can spread by word of mouth. African Americans are no strangers to the use of lies to stir up whites against blacks. At the dawn of the 20th century and deep into its day culminating around 1919 more than 4000 African people were lynched at the hands of white mobs. In many cases these mobs had been agitated by fake news, lies, that said black men had raped white women. One of the worst events was the bombing of "Black Wall Street" a prosperous area of Tulsa, Oklahoma, after a report that a black elevator operator had raped a white girl. It was later proved to be false but by the time the truth was reported there had been many blacks and whites killed and the black businesses had been bombed from the air.

Fake news is not new; it is old, but if a demagogue cannot control the legitimate media he must condemn it while claiming that he has the only true news. The demagogue seeks to isolate the masses from other influences and is frustrated when his word is contested by other sources. The challenge is problematic for the demagogue because he never wants to engage in logical debate over policy; he would rather dictate policy.

The Danger of Demagoguery and Lies

During the European medieval period in 1475 it is said that a false report went out in Trent that Jews had murdered a young Christian child named Simonino.[27] The report created such hysteria that many Jews were arrested and fifteen were burned alive. Pope Sixtus tried to quell the story but it had already infiltrated the outer reaches of the city. This so-called "blood libel" story was frequently the cause of pogroms in Europe. People were eager to hear that the Jews drank the blood of Christian children as an excuse to persecute and murder innocents. Consequently, students of human behavior have often spoken on the fickleness of stability in societies. As a student of communication and later head of the communication department at SUNY Buffalo I came to learn that powerful, incisive, and emotional rhetoric in the hands of a demagogue or despot could raze cities, kill the elder, murder religious leaders, and draw mobs willing to commit the most awful crimes.

Journalistic ethics evolved in Europe during the period after the invention of the printing press. Although the demand for references and citations increased after the trial of Galileo in 1610 the printing press in 1439 had already given historians a new instrument for crosschecking. News and information were considered so important that purveyors of fake news were often banned and some times fined for their lies. One of the most famous cases was that of the Dutch, Gerard Lodewijk van der Macht, who was banned four times by the Dutch until he eventually left for America. Some historians have claimed that Benjamin Franklin wrote false stories about "murderous scalping Indians" in service to King George

III to sway the public toward supporting the American Revolution. In my doctoral dissertation, "Samuel Adams and the Rhetoric of Revolution" at the University of California, Los Angeles, I examined the essays written by Adams in the Boston Gazette and discovered that the essays were often over the top in terms of reasons for a revolution. Indeed, there was hardly any real reason for the colonists to resist Britain since they were not taxed as hard as the people in Great Britain and had much more liberty than the average person who remained in Britain.

Fake news has been an instrument of some of the worst tragedies in human history. It seems that it goes with politics in a democracy and hence must always call into question the safeguards that we have in preventing absolute chaos. Consider that Marie Antoinette was beheaded because her face had been printed on a canard in 1793o,

In an effort to teach the masses how to recognize fake news the International Federation of Library Associations and Institutions published a list of criteria which should be used to assess information. Specifically, the IFLA says the public should:

1. Consider the source (to understand its mission and purpose)
2. Read beyond the headline (to understand the whole story)
3. Check the authors (to see if they are real and credible)
4. Assess the supporting sources (to ensure they support the claims)
5. Check the date of publication (to see if the story is relevant and up to date)
6. Ask if it is a joke (to determine if it is meant to be satire)
7. Review your own biases (to see if they are affecting your judgment)
8. Ask experts (to get confirmation from independent people with knowledge).

I am a believer in media literacy and think it will be a good idea for departments of Communication and general education in the United States to teach fact-checking skills to students and ordinary citizens as a way to train the masses in vetting processes for journalism.

Chapter 5

The Promise of the People

Demagogues have been so plentiful throughout the world that at this time they have established very obvious patterns, tactics and forms of communication. There are many methods by which demagogues have manipulated and incited crowds throughout history. No one demagogue uses them all, and no two demagogues use exactly the same methods to gain popularity and loyalty. Even ordinary politicians use some of these techniques from time to time; a politician who failed to stir emotions at all would have little hope of being elected. What these techniques have in common, and what distinguishes demagogues' use of them, is the consistent use of tactics to shut down reasoned deliberation by stirring up overwhelming passion in people. Trump's campaign rallies were often cries for blood, the throwing, so to speak, of red meat, to followers who barked for more. He basked in his followers' love of his often vile rhetoric of hate.

Sometimes, a genuinely wise politician, one who seeks not his or her own aggrandizement but work for the good of the people may speak in ways that represent demagogic rhetoric in order to "fight fire with fire." A statesperson usually is calm, cool, and calculating, but a demagogue is always exciting and emotional even when the tone is quiet. There is danger in a politician succumbing to the methods of the demagogue; one can lose a sense of perspective and moral compass. Once a politician has decided to "out demagogue" another politician he or she is moral trouble. I remember when the Alabama Governor George Wallace said that he once lost an election because his opponent "outsegged" him, meaning that the opponent went to the well of racism and segregation much more than Wallace did. His response was that he would never allow another candidate to "out-segg" him. In fact, both Spiro Agnew or Maryland, who went on to become Vice President under Nixon, and Winthrop Roosevelt of Arkansas, claimed that their opponents had "outsegged" them. This word meant that the white politician who appealed to the segregationist element among white voters in the South had a better chance of winning in much of the 20th century. Trump harks back to that day by trying to insure that no other candidate go to the bottom of the barrel quicker and more often than he does. This is his sad expectation in a deeply racist nation.

He who throws down the gauntlet first establishes the position for the opposition. But the person who takes up the gauntlet understands that defiance is necessary to defend personal and sometimes collective integrity. This is why the people cannot allow any demagogue to assume that they are weak, insignificant, or waiting for a savior. In a democracy, the people must pick up the gauntlet thrown down by a demagogue and run toward confrontation with their shields of truth, dignity, and self sacrifice.

Propaganda and Demagoguery

After the First World War when propaganda became a major part of warfare Edward Filene in 1937 created the Institute of Propaganda Analysis with the idea of educating the American public about propaganda techniques. Filene and his colleagues identified seven strategies used by propagandists. These were (1) Name Calling, (2) Glittering Generalities, (3) Transfer, (4) Testimonial, (5) Plain Folks, (6) Bandwagon, and (7) Card Stacking.

These propaganda techniques can be described in the following ways:

Name Calling

This means that the propagandists use negative or discriminatory terms to arouse prejudice or suspicion.

Glittering Generalities

In the use of glittering generalities the propagandists employs slogans or simple catchphrases with themes of honor, love, dignity, peace, family values, freedom, and patriotism.

Transfer

This technique associates a beloved symbol the flag to win popular approval.

Testimonial

This means that the propagandist employs testimonials from authority figures or celebrities to make an association with a cause.

Plain Folks

The idea here is to convince the people that the person is like them and shares their issues and concerns.

Bandwagon

The propagandist uses this technique to play on the human desire to be a part of a group, a crowd, or a important team.

Card Stacking

In using the card stacking technique the propagandist can cherry pick the arguments to make it seem that the propagandist way is the only correct one.

The preceding techniques are associated with propaganda but as you can see a skilled demagogue such as Donald Trump employs these tactics in offensive moves against those he considers opponents. Trump has used every one of the propaganda techniques during his presidency. But perhaps more to the point he has employed the techniques most associated with demagogic speech. These techniques have been developed over the past fifty years or more and have been based on studying the models of Adolf Hitler's Propaganda Ministry and Joseph McCarthy's House Un-American Activities Committee.

Thus, the main techniques are *scapegoating, fearmongering, lying, emotional oratory and personal charisma, accusation of weakness, promising the impossible, violence and physical intimidation, personal insults, vulgarity, ordinary folk actions*, and *simplification of ideas*.

Scapegoating

The most fundamental demagogic technique is probably scapegoating: blaming a group's economic and social problems on an out-group, usually of a different race, religion, or social class. For example, Senator Joseph Raymond McCarthy (R-Wisconsin) claimed that all of the problems of the U.S. resulted from "communist subversion." Denis Kearney, known for is violently anti-Chinese rhetoric and racial agitation, blamed all the problems of white laborers in California on Chinese immigrant.[28] Hitler blamed Jews and Roma for Germany's defeat in the First International European War as well as for the economic troubles that came afterward. This was central to his appeal: many people said that the only reason they liked Hitler was because he was against the Jews. Fixing blame on the Jews gave Hitler a way to intensify his sense of personal messianism; he would save the Germans from the Jews.

The rhetorical claims made about the scapegoated class are trans-generational, transcontinental, and transcultural. This appears to be the case even if the demagogue and the scapegoated group have little interaction with each other. It seems that the idea is to have a blamed group. Demagogues can have serial scapegoats; when one group is no longer effective in raising the opposition the demagogue looks for another group to attack.

In the American society the group most susceptible to demagogues appear to be white Protestant Americans. They tend to see themselves as the "We" people while blacks, Jews, Mexicans are viewed as the others. It is impossible for the scapegoated to be moved to "We" because the others are responsible always for the troubles of the "We."

The demagogue is keen to explain to his followers that the government has a "deep state" that must be cleaned out for him to exercise his authority to support them. He cheers on his followers with ideas that the Democrats are plotting against him and that all investigations into his personal finances, obstruction of legal processes, and collusion with Russians to undermine the campaign of Hillary Clinton are conspiracies against him.

Demagogues sometimes prefer to use the pronoun "they" to speak of the unnamed opposition. All "theys" are sexual predators, rapists, gangsters, members of M-13, and dangerous killers. This is the rhetoric that stirs up the demagogue's followers who believe that their women would be raped and that their lives would be in danger if the nation allowed immigrants into the country. Doom is sure to overcome the nation if the people do not give the demagogue more power and more adulation. The solution to this is to deny the demagogue the oxygen of admiration for unethical, trifling, and irrational behavior.

Fearmongering

Many demagogues have risen to power by spreading information and misinformation that is intended to frighten audiences. Donald Trump went directly for one of the most obvious and sick attacks by warning people that the "Mexicans are rapists."

African Americans have seen this story before. Jews know this tactic. Any people who have been labeled the "other" know the demagogue's cry that "they will rape your daughters" is one of the demagogue's most used tactics against African American males.

The American demagogue, Pitchfork Ben Tillman, at the turn of the 20th century often evoked colorful and pathetic racist imagery describing in vivid detail black men lurking to rape white women. In Tillman's rhetoric the black man had an innate weakness of character that meant he had to be kept from the white woman. Tillman who was elected governor of South Carolina in 1890, and elected senator repeatedly from

1895–1918, is one of the most virulent purveyors of demagogic hatred against African Americans.

Donald Trump may have never known Pitchfork Ben Tillman but his techniques, especially against Mexicans, is not far from Tillman's thinking about Africans.

Lying

The demagogue uses the lie as an effective instrument of people control. It is not just errors in policy or judgment but the absolute use of mendacity to influence the minds of his followers. Creating a web of lies to appeal to the emotions of his audiences seem to be a part of Trump modus operandi. Lies create additional chaos to a demagogue's administration. Sometimes the lies of the demagogue are unexpected and often trap the liar in more chaos. He usually demonstrates little regard for truth so long as he feels that his lies or those of his closest supporters benefit him politically. At some point the bizarre maze of lies captures the demagogue in an endless game of explanations.

My assessment is that Trump is opportunistic. He will do whatever it takes to arouse emotions and win favor with audiences. If the people are thought to have a problem with any person or ideas, Trump will identify with their sentiments in an incessant chain of lies. It seems that Trump is a compulsive liar seeking only to win friends, votes, and allegiances to him personally no matter the facts. When you are a demagogue who carefully monitors Twitter or Fox News for the most populist ideas you are likely to create your own reality to the extent that the lies will support your personal glory.

In Trump's case, he was much like other American demagogues who used the big lie, the lie to the side, the lie on other people, and the informational lie. The big lie, a Hitler favorite, is used to express something that is so wildly incredulous that people will say it must be so because who would be so foolish as to say that if it were not true. The lie to the side is the one that supports the big lie such as when Trump in explaining that "Mexico will pay for the wall" claimed that "in one way or another they will pay."

He slips easily from the frontal lie to a smaller version of the big lie. The lie on other people is when the liar claims that someone did something or said something or did not say something or do something. Trump argued that former FBI director James Comey was not told to go easy on Michael Flynn. The informational lie is used to confuse the audience with false data. When the demagogue wants to maintain a certain position that contrary to what is known he then tells a lie to establish a different number, set of facts, or reality.

One of the most classic cases of lying with impunity was when Senator Joe McCarthy of Wisconsin during the 1950s claimed to have "here in my hand" a list of 205 members of the Communist Party working in the Department of State. Not long after this claim, McCarthy decided that there were 57 "card-carrying Communists." [29]

McCarthy's challenge to the legislative branch of government was one of the most serious to confront the senate. His colleagues pressed him to "give us the names" and then McCarthy told them that the records were unavailable to him but he knew for a fact that there were "absolutely" 300 Communists who were certified to the Secretary of State for discharge but only 80 were actually fired. McCarthy changed again and said that he had a real list of 81 Communists that he would reveal later. This was all bluff, lies, and in the end he never found one Communist in the Department of State.

The Trump Playbook

Donald Trump was never a politician before he ran for the presidency. He was in construction of major buildings throughout the world, but mostly in Manhattan. His claim to fame was based on the broad celebrity he found in running beauty pageants, acting in *The Apprentice,* and his co-authored book, *The Art of the Deal*, where he offered his opinions about negotiations. Seeing him in the Office of the President one understands that a lot of what he called negotiation was about lying. No president has ever been accused of as many lies as Donald Trump.

The *Washington Post* says, President Trump lied 2,000 times during his first year as president. One of the biggest lies was that "Mexico will pay for the wall" that he wanted to build on the border. Mexico immediately denied this lie but he continued to use it to rally those who were ready for a hero. Lack of global knowledge, ignorance of politics, and willingness to believe someone who appears successful will get you a demagogue to play on your emotions while undermining the values of the nation.

Databases, centers, and institutes that categorize and track suspect political statements made by Donald Trump believe that through his Twitter account and public utterances Donald Trump makes an average of at least five false claims a day. Trump appears to claim a variety of exaggerated and false claims when it comes to his position vis-à-vis other presidents. He loves to claim that he has done something that "no other president" has done, and most of the time the claims are false. However, a demagogue knows that audiences do not have ways to immediately fact check the information presented as correct. Some even proclaim, "The president does not need to lie because he has all of the information." Such trust in a demagogue is dangerous. A demagogue wins the trust of followers with lies like:

- "We can build the wall in one year and we can build it for much less money than what they're talking about."
- On immigration policy, in the diversity visa lottery, "what's in their hands are the worst of the worst but they put people in that they don't want into a lottery and the United States takes those people."
- "We have tremendous numbers of people and drugs pouring into our country. So in order to secure the border we need a wall."

Trump's lies poured forth as if they were factual and authoritative condemning his listeners to inaccurate and confusing information. Like most demagogues, Trump realizes that the less educated masses can be manipulated because they have no systematic understanding of reality. They are fact-dependent upon opinion leaders and are apt to believe that rich people are smarter than other people, that white people are being deprived of their rights by Africans and Mexicans, and that the coal industry or shoe making factories will return to the United States. With little or no understanding of economics, the changing nature of world powers, or global conditions of trade, Trump's followers still believe that the United States can and should dictate the terms of global markets and regional power blocs. That time has long past and the demagogue is able to appeal to his followers with the lies that he will make it happen again because he knows that the most ardent conservatives actually believe that his vision is possible, even if he does not believe in it.

This situation reminds me of John Moody, the Fox News Executive Editor and Executive Vice President who wrote an Op-Ed in February 10, 2018 complaining that the American Winter Olympic Team of 244 people needed to be less "darker, gayer, or different." I guess the *darker* referred to the 10 blacks on the team, *gayer* referred to the two openly gay athletes, and I will not even proffer a guess what he meant by the *different*.

Stupidity has no ethnicity but it often seems quite popular among the ruling class white conservatives who sell a menu of conspiracies. Can you really imagine what emanates from an organization whose leadership practices bad racial thinking? The devaluation of rational thought suggests that instead of trying to close the Corporation for Public Broadcasting, the National Endowment for the Humanities and the National Endowment for the Arts, the Trump Administration should have sought to raise the level of understanding in the nation. Democracies are loss when the citizenry loses faith in the words and pronouncements of the leaders of society. The number one leader in the United States in 2018 is President Trump, but to say that he is a liar gives others with such proclivities the right to express their own willingness to lie. Mendacious behavior is like snow, so lies can color everything, but like snow they will melt if the sun is hot enough. When a racist New York City attorney named Aaron Schlossberg was accused of discrimination and yelling at immigrants who did not look like him or speak English as their first language he was just channel ing the Trump phenomenon against people who were not European.

Take the *Washington Post* article on January 10, 2018, that said, "Trump's claim about drug smuggling and the wall has been repeated 17 times, and by June, he had repeated it even more. With no proof Trump argued that building the wall would keep drugs from pouring into the country. In just two months into his administration, Trump falsely described the immigration diversity lottery more than ten times. At the very beginning of his presidential campaign, Trump demagogued the wall and low-balled the cost of building such a monumental monstrosity between Mexico and the United States.

Presidential Envy

However, given Trump's enmity or envy of former president Barack Obama, Trump's most repeated claim was that the Affordable Care Act was dying or essentially dead when in fact the Congressional Budget Office was saying that the Obamacare exchanges were not imploding and remained stable with healthy enrollment. But Trump's assault on the Obamacare program was relentless until the legislators failed to repeal it. Regardless to how hobbled it had become because of the attempt on Trump's part to kill the mandate that every person should have health insurance, people still signed up for it at a faster rate than the year before Trump's administration took office.

Given Trump's antipathy toward Obama one wonders why he sought to repeatedly take credit for political, business, and social events that were started under the Obama Administration. Many business decisions were made prior to his election and yet he took advantage of the business actions of giant corporations. Several economic achievements were set in place by the Obama's Administration's rescue of key technological industries in the nation, especially the automobile industry, and the shoring up of the financial institutions that were teetering on absolute failure. One can question, and people have, the idea that there were financial companies "too big" to fail. Nevertheless, the saving of the economy from the depression of 2008-09 must be attributed to the quick and decisive actions of the Obama government.

One does not have to revisit Obama era policies to see the inconsistencies in the demagogue's rhetoric of simple lies. When Trump's tax law was passed in 2017 it was demagogued when Trump claimed falsely that it was the largest tax cut in history and that the United States had the highest tax rates in the world. In January 30, 2018 CNN Politics reported that Trump overstated the facts with his boast. CNN researchers said, "But arguably, President Barack Obama passed a larger tax cut by making most of President George W. Bush's cuts permanent. President Ronald Reagan definitely did. And Presidents John F. Kennedy and Lyndon Johnson probably did, too." Furthermore, CNN Politics said, "Tax analysts have been unequivocal that Trump's claim is not true." This means that Trump lied.

I have concluded that Trump is adding a flavor to American demagoguery that we have not seen much in this relatively new century. The obvious strategy is to tell many lies in a short period of time like one shooting a shotgun with many pellets flying in a wide swath with a big bang. Ordinary people do not know what to make of this: Are all of these lies? They surmise, "It cannot be that everything the President says is a lie, can it?" As a master of demagoguery Trump also knows that the media, despite its attempt, will not be able to fact-check everything he says in real time. The lie lives with two legs for a while before one or both legs are destroyed by truth. This is a game that should never be played in a democracy, but is always played by selfish politicians for personal political gain.

Emotional Oratory

Many demagogues have demonstrated remarkable skill at moving audiences to great emotional depths and heights during a single speech. Sometimes this is due to exceptional verbal eloquence, sometimes to personal charisma, and sometimes both. Hitler demonstrated both in his speeches to German audiences. Accordingly stories abounded that Hitler's eyes hypnotized people, that they became immobilized because of his personal passion, and that he overwhelmed his audiences with his ability to voice what they wanted to say but feared to say. What demagogues do best is to allow people to believe in them as the voice that they want. From the most bitter racism and unspoken hatreds to the colorful language that restrains them the reactionary masses are able to pile onto the shoulders of the demagogue their dislikes, preferences, and foul language. They become like infants to the parent.

Communication scholars such as Fred Casmir and Erika Vora, as well as others, have categorized the rhetorical style of Hitler. He would began his talks slowly, pulling the people in with a slow style of speech that was almost boring until he raised the resonance, telling about his life as a child, and then he would go into his blame game. Well, Trump did not come from poverty and he does not have that story to tell yet in many of his speeches he minimizes the nature of his upbringing and he emphasizes the idea that he had to work very hard. Rarely does he mention the fact that when you start off in a family with millions left to you there is an advantage unlike that of the masses of Americans. He did not suffer humiliation that often comes with the lack of food or resources; Trump lived the life of a rich child who probably bullied other children who had less considering his current personality behaviors. There is certainly nothing wrong with having been lucky enough to have successful parents, but there is something terribly wrong with trying to identify that fact as similar to the life of the coal miners of West Virginia or the potato farmers of Idaho. The problem with Trump is not his money or his silver spooned upbringing but his utter lack of empathy with the millions who yearn for enough resources to feed their families, to pay for health care, or to cross a border into a country that would give them a chance to succeed. Rather than support the human desires, wishes, and needs of the masses, Trump seeks ways to exploit the people.

Hitler shrieked hatred of Bolsheviks, Jews, Roma, Czechs, Poles, or whomever he claimed stood in his way. Like Trump's modern rhetoric of mocking the disabled, criticizing those he thought were too fat, ridiculing those he considered too weak or too slow, and so forth, Hitler mastered the technique of ridiculing and insulting other people, sometimes telling his audiences that he would destroy such people. How could people we think of as reasonable and normal, maybe even religious and pious, accept such vile and poisonous rapport with a demagogue?

I suggest that Trump like Hitler was a student of human nature and in Trump's case his time in military school, his practice as a young bully, his brash and pompous behavior as a young man, and his adult life as a sexual predator according to his own words confirmed in his heart his ability to manipulate people through bluff.

Trump is not a natural orator; his appeal is based in his populist promises to an energetic minority that he would save them from the *others*. One does not have to master metaphors, similes, periods are any other elements of style to move those who seek relief from what they consider to be existential threats. Apparently what one has to do is to give people the most wildly irrational promises and then pound that into their heads, even as a conspiracy theory, to the extent that they come to believe it more than the speaker, and you will have a following. We have seen this narrative with religious fanatics who claim the most irrational ideas with passion and then find other people willing to believe even to entering the Brand Davidian or James Warren Jones' Jonestown cults until their deaths.

Enter Race as Tactic

What history teaches us is that race often plays a fundamental part of the American demagogue's bag of tricks. In using oppositional races rather than uniting of races Trump harks back to a nativist white American emotion. America's history is replete with anti-African demagogues such as James Kimble Vardaman, Governor of Mississippi from 1904-1908 and Senator from Mississippi from 1913 to 1919. When Theodore Roosevelt invited African Americans to a White House reception, Vardaman railed "Let Teddy take coons to the White House. I should not care if the walls of the ancient edifice should become so saturated with the effluvia from the rancid carcasses that a Chinch bug would have to crawl upon the dome to avoid asphyxiation." Like most demagogues, Vardaman's speeches were often devoid of serious content as he preferred ceremonial talk even when the occasion called for deliberation and argument. His speeches were avenues for him to drive home his irrepressible hatred of African people.

The demagogues' appeal and deeply felt emotions have been known to win elections for them. They could even have opposition from the free press, seeking to tell the truth, and yet they are able to attract audiences that believe the lies, distractions, assaults, and broken promises peddled by the demagogue who in most cases seeks to create an autocracy.

Even as president Trump drowns out the media with Twitter to his followers. They experience his tweets as the words from the autocrat who denies them, within the context of his own universe, any other information. They wake up in the morning and go to bed at night intoxicated by the demagogue's prevarications. In a sly and deceitful twist to assaulting the media experts Trump claimed that the mainstream media, all of the major traditional news agencies, did nothing but push "fake news" and the only true news was advanced by those who praised Trump.

It appears that the demagogue does not care about any fact even if the fact is known by millions of people; he is able to twist the information to support his reasoning. The only politics of the demagogue is the saving of himself. This is why it is easy for him to say one thing today and another at night. He is willing to use information, reveal any sources, conceal any incident, weaponize classified information to protect his vanity.

Congress did not limit his abuses of power during the first year of Trump's administration. Legislators were weak and ineffective in stopping Trump from a steady diet of self- aggrandizement. Republicans, in leadership of the House and Senate, seemed paralyzed or frightened nearly to death to take on Trump.

What the American people witnessed with the election in 2016 of Donald Trump was the capture of state power by a radical clique willing to destroy public institutions in education, justice, energy, health, and other sectors in order to create an autocracy where the "leader" can be glorified by marching bands and demonstrations of military armament. Steve Bannon, the Breitbart entrepreneurial white nationalist strategist, who would later leave the administration and go to France to peddle his racist ideas to the Le Pen crowd, was to be the Grand Master of Trump's rise to glory as the leader. Alas, it was not to be because demagogues must always seem to have their own minds and some people thought and wrote that Bannon was Trump's mind.

In fifth century BC Greece, Cleon, like many demagogues who came after him, believed that to keep the people's loyalty he had to demonstrate brutality as a measure of his power. For the demagogue, talking, negotiating and mediating show weakness that compromises the demagogue's persona of powerfulness. Cleon believed that people despised those who treat them well and fear those who show brutality. While demagogues like to make strength especially brute force one of their weapons to fight against enemies the logic does not hold in human communication. Most people love those who love them and despise those who despise them.

General Cleon opposed the idea of debating the issue of recalling ships that he had sent the day before toward Mytilene to slaughter the population. He saw debate as a sign of weak government. In fact, debating issues were considered the preoccupation of the idle and weak and as for him, the only reaction to enemies was to lock them up or to kill them, In this way, he argued that the people would see his strength. Through the centuries there have been variations of this thinking despite the obvious truth that people prefer governments and governors who tell the truth, respect their person and their rights, and seek the higher ideals of life.

One variation to destroying enemies physically is the destroying of their reputations. When Senator Joe McCarthy did not have evidence he persistently claimed or insinuated that those who opposed him were communist sympathizers. McCarthy's line went like this: "I am for the American people and the American people are against communism. I am against communism, and you are against me, so you must be a communist."

Trump used the demagogic technique of promising something so wild that he had no intentions of really seeing it through, but rather using it as a technique to arouse emotions. Should it become law in some form or another he would claim it but the emotional effect is all he was looking for in making the promise. This is why he had no idea of what it would cost or look like or which of his building contractor friends would revel in receiving a contract to build the wall between the United States and Mexico. Most demagogues give empty promises that they do not intend to fulfill.

Governor Huey Long of Louisiana was also one of the prominent demagogues of the 20th century. He once promised during a campaign that if the people elected him president he would give everyone a house, a car, a radio, and $2000 annually. He had no plan for doing this yet thousands of people joined up with him in a Share-the-Wealth campaign as he sought to become president. Although never thought of as a rabid racist such as his son, Russell Long, would turn out to be, Huey Long nevertheless was a southerner with proclivities and attitudes on race based on the prohibitions placed on blacks in Louisiana by the 1898 Constitution.

Trump obviously believed that he could promise anything because he felt that psychologically he knew that people would buy anything if you said it long enough and loud enough. They would even buy and do things against their own political and economic interest. Using his prior business success Trump became, in his mind, the one who would solve all problems better than any previous president. I think that there is a certain callousness and cruelty to the exploitation of working class people, miners, farmers, and ordinary citizens when they are given impossible promises.

During his campaign for the nomination and then the presidency, Trump encouraged his supporters to violently intimidate his opponents. This is a demagogic technique used to encourage loyalty and to discourage from opposing him.. Pitchfork Ben Tillman used this technique of intimidation very successfully in South Carolina and he was repeatedly sent to the Senate. Can you imagine that he actually spoke in support of lynch mobs organized against blacks and supported a South Carolina Constitution of 1895 that effectively prevented blacks from voting? Nothing is far-fetched when you think of the motivations of a demagogue. They will do or say anything to get elected.

Trump, the American modern demagogue, encouraged physical intimidation to demonstrate his strength to the masses at his rallies. He would provoke the hecklers and demonstrators for the entertainment and arousal of his followers who on occasion assaulted some of the anti-Trump people. All of this theater played into the hands of the demagogue who told his supporters that they had to defend him because he had these people who wanted to prevent him from being elected although he knew that the elections were rigged.

Trump does not have the ability to make a discursive argument or a long discourse on any policy issue. He is a tweeter who apparently has a short span of attention and seeks other ways of conveying messages rather than policy statements. Trump uses insult and ridicule as a way to shorten deliberation. If he does not use narcissistic praise in a meeting, then he uses ridicule of people, plans, and policies, There is no discussion in detail of what the plans or policies mean to the people. Trump is rather about the show and not about the policy; in effect, policy is made by tweets. Heis the first demagogue to use tweets to keep his followers attuned to his antics. At his rallies with mostly unsophisticated audiences he is able to shout down reasoned argument with assaults, threats, and intimidation or ridicule and shaming. The infamous "Pitchfork Ben" Tillman, for example, got his nickname from a speech in which he called President Grover Cleveland nothing but "an old bag of beef" which he would bring a pitchfork to Washington to poke him in his old fat ribs. The people laughed whenever he gave that line and soon they called him by the name "Pitchfork Ben."

Trump has mastered the vulgarity and vileness of the lowest type of person in his public and private conversations. Devoid of real passion for truth, goodness, and justice, he practices a feigned caring by resorting to platitudes and sentimentalities in the presence of the gravest situations. Name calling as we have seen is one of his favorite habits. He reminds me of the reports of James Kimble Vardaman, the skilled demagogue, who once called President Franklin Roosevelt "a coon- flavored miscegenationist"[30] What could have compelled Vardaman to post an advertisement for "Sixteen big, fat, mellow, rancid coons" to sleep in the bed with Roosevelt while he visited Mississippi? His antipathy against blacks was known by all Mississippians, but any white who was fair and just toward African Americans came under his glare and fire.

Trump's demagoguery in the face of immigration is similar to Vardaman's in regards to black people. Any progressive white politician, in Trump's narrow classification, a Democrat or a maverick Republican, would be attacked. As we know Trump enjoyed the technique of creating insulting epithets for opponents as a way to keep the public from considering serious matters by diverting their attention to laughter. Personal invective against an opponent may help gain a few points for humor but does nothing to create rational discourse.

In many ways Trump has shown a contempt for women, making them the object of ridicule, calling them names if they accused him of sexual harassment, and denying that he or his friends ever assaulted women. The dignity of women is not in his vocabulary. The verbal assault against Congresswoman Fredericka Williams by Trump and his Chief of Staff John Kelly was one of the most despicable examples of a demagogue's bag of tricks. They insulted her by telling a lie and even after the news media had revealed a video that showed it was a lie they held to the lie and never apologized to the congresswoman. This was so blatant and so bold as to cause even those who tried to protect Trump's administration to encourage that they retract their statements and apologize to the congresswoman. Of course, here we see that a liar who holds onto a lie when they have been shown it to be a lie is neither iconic nor peculiar; this liar is of the same corrupt substance and essence as the worst American demagogues.

The Demagogue's Protocol

I think that what we have seen with Trump is that he is a disrupter, but he is not a builder. He destroys rules and without rules you cannot create any thing of substance. Trump has violated the basic protocols of diplomacy, collegiality, government, and personal style. He insulted a Gold Star family, that is, a family that had lost a son to battle who had fought as an American in war. He threatened to jail his political rivals, encouraged his followers to expect violence and to be prepared to dish out violence as well. He once pointedly told Hillary Clinton that if he became president he would get "officers to look into your situation." A third rate threat on an opponent should be beneath American politicians, but what has happen is that Trump as a demagogue has often seen security officers and other government officials as serving him personally rather than serving the nation. He violates all protocol to go after his opponents and critics in a negative way by insisting that the officers of government should serve him. This is the radical position in which the country finds itself as Trump signals the decline of discourse and the ascendancy of bombast in political rhetoric.

How Democracies Die

Democracies are placed on the path toward death when the political leaders are corrupt and incompetent. Trump's promise to hire "the best people" turned into a cadre of losers such as Michael Flynn, Paul Manafort, Rob Porter, John Kelly, Carter Page, Rick Gates, George Papadopoulos and nearly 130 people serving the White House without being able to pass full security clearing after more than a year after Trump's election.

Western democracies probably learned decorum and protocol from the French professional civil class. One could of course trace the origin of *respect* in government to the ancient African concept of *maat* that meant harmony, order, and balance within the context of dignity-affirming words.[31] Trump has violated many of the most sober standards established by the long history of democracies. The aim of standards, rules of political speech, and good manners is to create conditions for less emotionalism and more deliberation. However, when you have individuals who are not readers of theory, information, or literature, you tend to get the pedestrian vulgarities of Trump's athletic clubhouse.

Trump's vile language most likely comes from vulgar thoughts. He claims that his talk is the way "men talk in the locker room" but he lies; it is the way he talks wherever he is because he attacks all standards. While he may think that he is thumbing his nose at the traditional politicians, he is really demonstrating that he is ignorant of the minimum standards of presidential decorum. Barack Obama, for example, elevated the office by seeking to express the elegance that should flow from the top person in the United States government.

The problem with Trump's lack of integrity is that he could not help himself. He lied and placed around him others who would lie for him and give him the praise that he craved. Whatever length it took him to go for praise, he would travel that distance to make himself the dominant personality in the room. He would lie often and without any reservations. Like other demagogues Trump mastered the technique of the audacious act. He would shock his rallies with outlandish lies, tell reporters lies about Barack Obama wire tapping his home, and lying about ordinary things regarding his financial condition or his tax filings. Trump is a man without restraint who is shameless in the search for attention from all sectors of the society. Because he had been in media he knew how to get press coverage and he manipulated the hungry-for-news or spectacle seeking reporters, feeding them stories with dubious reasoning. He became a metrics man, measuring his numbers against those of his opponents. This is why he had his media people to lie that his inauguration crowd was the largest ever, which it was not. T

Aristotle is said to have written that the demagogue Cleon shouted on the public platform and used abusive language. When I read Aristotle's description of Cleon it was a further indication that demagogues have neither style nor substance and do not change through the years.

Down to Earth

Even with his billions Trump works on the emotions of his audiences using the plural pronoun "we" as if he is a part of the same group as his followers. This is a common demagogue tactic. By using this type of language Trump drags his followers into the circle of villainy that is a part of his career.

It reminds me of Georgia demagogue governor Eugene Talmadge's decoration of the Executive Mansion grounds with a barn and a henhouse because he was just like his voters. Trump tells the West Virginia coal miners that he is going to bring back their old mines. This slick Manhattan businessman convinces many ordinary people that he was one of them. Like Talmadge, Trump had speeches for different groups of people. He often spoke to veterans and farmers as if he could do for them, and would do for them what others could not or would not do. Talmadge spoke to his ordinary people with followers by railing against "nigger-lovin' furriners".

Donald Trump's response to the modern world is the most backward of any recent president. He is anti-immigrant, anti-Muslim, anti-science, anti-technology, and anti-knowledge. Having such attitude means that he is capable of making complex issues appear quite simple. Whether this is a matter of his own incompetence or a stroke of genius in bending the complexities of climate science, for example, into a simple type of Trumpism like "I think we have a cold winter, there is no climate warming."

Many political, technological or social issues require patient and serious contemplation which means definition and analysis before action. The assertion that anything that poses a problem to American democracy can be negotiated with some notion of the art of the deal is quite sophomoric. Such oversimplification of complex problems imperils the nation because it demonstrates the naiveté of the president.

There are several rules followed by all demagogues. In the first place, they have to attack the rule of law. Considering themselves beyond the law, outside of the political class, and holding mandates to change the rules, demagogues seek to make exceptions to rules that constrict and restrict their actions. Trump is no exception to the cases. He has sought to deal with his Chief of Staff as a creature of his own making and to put his Media Spokesperson in an intolerable position where that person has to lie in order to keep the position. In this way, the people in the media positions are given no choice but to defend a fuzzy line between that which is real and alternative realities. Breaking rules is often the results of not knowing the rules or refusing to accept the rules as applying to your actions. Trump likes to say "I hereby demand" as if he is asserting real power and showing that he is in charge, but his rhetoric rings hollow because of the many transgressions that he has made in the are of protocol. I am sure that as a demagogue he believes that it is unnecessary for him to study the history of the American institutions or the protocols of diplomacy and the ways of governing. You do not bark commands or demands at other citizens in a democracy; each person has rights that cannot be taken by the president, and presidents do not live above the law.

Trump's pattern is to claim that the ends justify the means. When he sets his mind to a goal, he will do anything to get that end. This means that he will lie and ridicule people to justify the means.

Trump's mentality is much like that of other demagogues, "Only I can fix it." The demagogue will not relinquish his attachment to his exceptional qualities.

Trump flip-flops on many ideas.

The Washington Post, of January 10, 2018, tracked the flip-flops of Trump. During the campaign of 2016 Trump said that the unemployment rate was really 42 percent and that official statistics were false, yet he later told his audiences that the unemployment was thelowest in 17 years, but the truth is that the rate was already 4.6, the lowest in a decade.

Shifting Positions

In an astonishing 91 times, Trump has celebrated the upward swing in the stock market during his administration yet during the presidential campaign he said as the market rose under Obama that it was a bubble that would soon burst as the Federal Reserve started raising interest rates. The Fed raised rates at least four times since Trump was elected and yet the stock market has not plunged. The rise in stock prices that began under President Barack Obama in 2009 has continued until 2018. Brutalizing all conventions of ethics from any tradition of religion or morality Trump refuses to apologize, admit that he was wrong, or to compliment the Obama administration. Instead he goes on the attack of his predecessor in office and condemns the media for not supporting the Trump agenda.

Attacking the News Media[

Since information from the legitimate press, where journalists double check information, can undermine a demagogue's spell over his or her followers, modern demagogues have often attacked it intemperately, calling for violence against newspapers who opposed them, claiming that the press was secretly in the service of moneyed interests or foreign powers, or claiming that leading newspapers were simply personally out to get them. Huey Long accused the New Orleans *Times–Picayune* and *Item* of being "bought", and had his bodyguards beat up their reporters. Oklahoma governor "Alfalfa Bill" Murray (1869–1956) once called for bombing the offices of the *Daily Oklahoman.* Senator Joe McCarthy declared that *The Christian Science Monitor, the New York Post, The New York Times, the New York Herald Tribune, The Washington Post, the St. Louis Post-Dispatch,* and many other leading American newspapers were "Communist smear sheets" under the control of the Moscow.

The central feature of the practice of demagoguery is persuasion by means of fear-mongering, shutting down reasoned deliberation and consideration of alternatives. Demagogues pander to raw emotions, racism, prejudices, bigotry, and unreason. Indeed "McCarthyism" was coined in 1950 to refer to a kind of reckless, unsubstantiated, smear type of rhetoric. All one has to do to understand Trump is to review the methods of persuasion used by most demagogues throughout history. Demagogues usually start with lying and then attacking the press. This is definitely the *modus operandi* under the Trump regime where if the demagogue cannot control the media, or play the media for fools, he will attack the press or simply use "alternative facts" on his own Twitter account.

McCain's Lone Republican Counter to Trump's attack on Media

"President Donald Trump's assault on the press, most often through barbed tweets and harsh words, has manifested itself more dangerously around the world," Sen. John McCain wrote in a *Washington Post* op-ed.[32] McCain went further to say that

"Reporters around the world face intimidation, threats of violence, harassment, persecution and sometimes even death as governments resort to brutal censorship to silence the truth."[33]

The attitude Trump displays toward the media is unworthy of a president of a free people whose freedom is rooted in the ability to express views contrary to those of the president. Thus, one cannot simply say that Trump is hypocritical when in fact it may be a case of an individual who has no sense of appreciation for other humans. For Trump, it appears that the only function he sees for others is to serve his own ego. I believe this is the reason Trump continues his unrelenting assaults on American journalists

Arizona senator McCain cited statistics from the Committee to Protect Journalists that 262 reporters were jailed in 2017. McCain also said that the CPJ's figure that 21 reporters were imprisoned last year on charges of "fake news," the phrase used by Trump to degrade reporting that he believes is against him.

The Special Effect of Demagogic Racism

The irrational issue with demagogues is that they distort logic and truth, and also stir up the masses and sow deep divisions in society with lies. The black and white thinking and hostile rhetoric toward the "other" that they encourage is damaging to the fabric of society and critical thought.

Nevertheless, demagogues have been a societal staple since ancient times. Wherever there are segments of society that can be riled up, there will be demagogues. Well-known demagogues include Huey P. Long, Joseph McCarthy, and Father Charles Coughlin, a precursor to modern talk radio hosts. A review of other historical figures that have earned the label of "demagogue" shows convincingly that Donald Trump is not an aberration in the American society although one hopes that he is an aberration in the presidency.

A Parade of Political Hooligans

Theodore Bilbo was Mississippi's most dominant personality during the first quarter of the 20th century. Bilbo was a segregationist with a fiery spirit and an extravagant attitude who believed that blacks were an idle and slothful people. There was nothing in Bilbo but hatred for Africans who he saw as subhuman. Bilbo's rudeness made him notorious for the lack of civility but he believed that the "general feeling" of whites was also with him. His contemporaries called him a strutting peacock although he was just 5'5".

As a demagogue Bilbo engaged in contempt for blacks whenever he could in order to gain credibility with poor whites. The use of the poor and uneducated white populations is a cardinal principle of American demagoguery. The idea is to invest the poor and uneducated whites with the powerful privilege of skin royalty by making them the inheritors of the dream of a white nation with others who purpose was to serve whiteness. Attractive as this dream might be to some whites, especially those who saw Trump as a sort of white knight, it was an impossible reality.

The United States of America embraced the vision of a nation seeking to enable all of its citizens and others with the democratic meme since the turn of the twentieth century. Of course, this was contrary to the political ideology adhered to by Bilbo and the nation found itself torn between race wars and racial reconciliation for decades. Yet pounded into the national rhetoric by Africans, women, and labor unions the idea of diversity took hold and became a leading tendency in the latter half of the twentieth century. This particular characterization of the nation gave birth to the progressive sentiments that challenged the neoliberal and conservative traditions that have always found their way into American discourse. Thus, Bilbo represented the reactive thrust, the element of radical nationalism based on race, class, and a search for the *status quo, ante*.

Bilbo said that his opponents were corrupters of Southern womanhood, and skunks who stole Gideon Bibles from hotel rooms. He told people that he knew that many communists were in the government and so the people should elect him to clean it up. This is reminiscent of Trump's promise to "drain the swamp."

Demagogues always seem to know more than other people because they lie more easily than others and do not have a penchant for research or facts. The fact that Bilbo went around speaking about communists in the government without ever citing any evidence opened the door for Joseph McCarthy to walk down the same stench-driven corridor nearly twenty-five years later. Trump would later take up the cry of the "deep state" being unable to speak about communists since the reports were that his government had extremely close, even criminal, contacts, with Putin's Russia.

When Bilbo got to the Senate he railed against Africans. He was a vicious attacker on blacks who introduced bizarre bills meant to resettle blacks into West Africa. He once demanded that all Africans should be moved from Washington. As head of the Senate Committee with oversight of the District of Columbia Bilbo sought to stop the growth of the blacks moving into the district. In fact, he was adamant that there were too many black people in the city and they had to move. Of course, the city's population prevailed and blacks remained in the city long after Bilbo had left DC.

Like all demagogues Bilbo knew that his vile language and coarse style would draw some followers. There was segregation in Washington in the 1940s and Bilbo made sure that the city did not forget that DC was a southern city despite his inability to evacuate all blacks.

Trump also borrows from Lewis Charles Levin, a crusader as he called himself, obsessed with the idea of getting rid of all aliens who were corrupting America. Immigration so often seen as a part of the genius of America and a pillar of the diversity that identified the United States as different from European nations became a rod to poke in the sides of the emotional body of Trump supporters. He defied the established virtue of seeking more diversity and more unique gifts that constituted America's special quality for a bitter anger at all immigration. Trump started his assault on the Mexicans and then on the Dreamers those who were brought to the United States by their parents and who were raised in the United States, served in the army, went to college, and worked as Americans. They were anathema to his white right ideology of narrowness. He rolled back the executive order that had been made by President Barack Obama providing a pathway to normalizing the residency of the Dreamers. Trump declared a March 2018 deadline for them to be out of the country although their only known home was the United States. Not only was this position irrational it was cruel and mean-spirited. Through no fault of their own the Dreamers lived in the United States and served their country, signed up to give their names and addresses to be legitimate only to find that Trump would exploit their information to try to expel them from their country.

Trump's demagoguery exposes his anti-immigration stand to be against all foreigners, not just the Dreamers. He has sought to change even "legal" immigration to limit the number of people coming from Africa and Asia and Latin America and the Caribbean. He shouted his anger at the immigration of Haitians and Africans from "shithouse nations" asking his Republican colleagues and aides in the presence of Democrats why couldn't America get "more Norwegians?"

One is apt to think that this type of behavior was uncharacteristic of a president but Trump has changed the equations of decency succumbing to the most negative aspects of the American personality: selfish, hegemony, racist, and willing to degrade any one who does not think or act like he does.

Lewis Charles Levin foreshadowed Donald Trump with his self-styled crusader persona by cursing at those who were considered aliens. His single idea was the protection of white protestant culture against the "danger of subversion by the influx of that horde of aliens, who combine to break down its barriers, that they may command in the citadel, or overrun the land." Perhaps those who give their support to Trump have forgotten that Irish, Italian and Greeks could not have held their places in Levin's cultural war. He did not care if they were legal or undocumented; he simply did not want them to defile the America he knew. For Trump, the idea is to keep out those who are not white and to bring only those who add to the white community. This is a demagoguery based on fear of the other.

Levin's declared enemy was Roman Catholicism that he believed to be an organization of "alien" European criminals.

Levin was the son of Jewish parents, born in Charleston, South Carolina, in 1808, at the height of white enslavement of Africans. The Constitution ended the importation of Africans to be enslaved in the year of his birth. However, the sentiment among the slaveholding South was anti-African. Levin grew up in this environment in South Carolina and would later attack immigrants form Europe. Some people believe that Levin felt some cultural alienation himself and this caused him to rant against the Roman Catholics. After he was wounded in a duel in the South, Levin settled in Philadelphia becoming the editor of *The Temperance Advocate.*

At the age of 51 he sold the *Temperance Advocate* and purchased the *Daily Sun* that he used to advance his obsession with the country's growing Irish Catholic population. A major conspiracy theorist Levin believed that the Roman Catholic Church was trying to conquer the United States by bringing more Irish to overwhelm the election process.

When Levin held a campaign meeting on May 3, 1844 in Kensington, just north of downtown Philadelphia, a mob gathered outside to attack him for his insulting remarks about Irish people. Levin's retaliated with their attacks and threatened to burn down St. Michael's Church. Subsequently, Levin's group, calling themselves the Nativists, held another rally three days later with his followers armed with guns. When an argument occurred between an Irishman and a Nativist it opened Philadelphia to several days of riots in which the Nativists sought to burn down every Catholic Church in the city. This anti-immigrant attitude was latent in the American society and it took the demagogue Levin to resurrect it with his vile rhetoric of hate.

The demagogue Levin rode the wave of anti-Catholicism all the way to being elected to Congress from 1845 to 1851. After the riots, Levin rode the wave of anti-Catholicism to election to Congress in which he served from 1845 to 1851. His speech to the House of Representatives in 1848 is one of the vilest anti-religious speeches ever made against Catholics in the Congress. His obsession with foreigners, especially Catholics, bordered on irrational fear and at the end of his life, some speculate that his bitterness may have driven him to insanity and death. What we see in the Trump era is that some of the people who were attacked in the past for being foreigners are now the same people, that is, their descendants who today attack others. How is it that Catholics, Italians, Jews, Poles, and Irish side with the older vile rhetoric produced by those who claim that the country belongs to them alone? Do they not remember their own history? This shows us that a demagogue can use any group as the out group and can get people to accept hatred against that "out group" with emotional rhetoric for political gain.

The Ubiquitous Nature of Demagogues

The United States is not the only country to have demagogues. Udo Voigt, the German politician, who served since 2014 in the European Parliament, once called Hitler "a great German statesman" causing consternation among many Europeans. Voigt thought that the Nazi, Rudolf Hess, should be nominated for the Nobel Peace Prize. Voigt's personal style might be called that of an individual shock trooper who sought to stir up the public with his wild pronouncements.

Most demagogues claim that they are the only ones with the answers for the masses, Voigt, for example, called for an armed uprising against the government of Germany in 1998 because he believed that the country was occupied by outsiders and anti-Germans and he was the only clear-eyed German to see the problem. Reaching out to the United States he made contact with white racists, including David Duke, the former Ku Klux Klan Wizard and anti-Semite. Duke called Voigt the "true chancellor of Germany."

Demagogues tell lies anytime and any place with no problem. Voigt, for instance, said in contradiction to historically accepted death count of six million Jews dying in the Holocaust that it was only 340,000. German racist, Udo Pastoers, taking Voigt's rhetoric as an example, was charged with inciting racial hatred, a crime in Germany, for calling Germany a "Jew Republic" and calling the Turkish minority "semen cannons."

Like Trump's rhetoric the German demagogues isolated groups of people that they thought should be denied something that others, mainly white Christians, were entitled to have.

Trump is morally corrupt but he is not politically inept because the lies that make him immoral have served to advance his cause with his followers. They support him as if he could tell no lies and serve as his brick wall of protection from the spotlight of reason.

I find it difficult to understand those who hear Trump's lie that he has been "tougher on Russia than Obama" and believe it given the facts that while Trump refused for more than a year to impose the sanctions voted by Congress on Russia, Obama actually expelled thirty-five Russians, took control of two of their properties, imposed sanctions on the country, and told President Putin to "cut it out" during the last months of his second term.

Trump's demagoguery borders on psychopathic reasoning. It reminds me of the logic finessed by the racists in 18th century Pennsylvania who wanted to deny free blacks the right to vote who argued in court that although those Africans seeking to vote were free, "being free negroes is not the same as free men"; therefore free blacks could not be allowed to vote because they were not free men. Racism creates not just distortions of truth, but the most bizarre contortions of reason.

Defending Process from Disruption

Trump's acrimonious rhetoric and chaotic administration underline his attacks on the democratic process. It is as if the president came into office off the streets with no obvious familiarity with the Constitution, political protocols, or consultative realities of the political process. Like a child who cries when he cannot get his way, Trump not only cries, but seeks to destroy what he cannot comprehend. The Trump phenomenon corrupts ordinary conversation and renders rational argument and discourse mute in preference for shouts of "You are fired!" It is as if Trump's doggedly pessimistic attitude toward truth, facts, and logic imperil every sector of society.

One sees boldness in the white supremacist community because Trump has made it popular among some who may have been latent racists. They want to be like him, to have money and power, and to dominate others with commands and demands of the oddest kind. This sickness is seen in the play of sports teams, on the playgrounds of urban communities, in the circles of wealth and status in major cities, where people dish out and receive insults as if these are calling cards of relationships.

Charlottesville was Trump's time to demonstrate leadership and he demonstrated his waffling style on the issues of bigotry. When neo-Nazis, white supremacists, and bigots marched in Charlottesville, Virginia to protest the removal of Confederate statues representing the supporters of rebellion against the Union and support for men who fought to preserve slavery, Trump choked and came up very short in his remarks. It was August 2017 and hundreds of chanting white supremacists yelled hateful words and one, James Alex Fields Jr. plowed his Dodge Challenger into a crowd of counter protesters killing Heather Heyer, 32, of Charlottesville, and wounding nineteen other people.

Trump said on August 16, 2017 among other things that there were good and bad people on both sides. A firestorm of protest against his comments erupted when he said to reporters, ""I think there is blame on both sides." How could a president be so detached from history and reality that he did not understand the moment? Perhaps he understood precisely the moment and was himself happy to roll back the nation's discourse to racial epithets and deep resentments.

Confronting the Demagogue

So what must be done? The antidote to demagoguery is always truth. I suggest that the American public has been served well by the mainstream media that has held Trump responsible for the lies that he has told from the comfort of the oval office, Mar-a-lago or his Manhattan Penthouse. Surrounded by white nationalists like Steve Bannon and Steve Miller during his first year, and liars for him like Sarah Huckabee Sanders, Kellyanne Conway, and Raj Shah, Trump has fired his cannons of lies every day in an effort to flood the airwaves with chaos. Only the Fourth Estate has challenged him in his own arena, the public platform. During Trump's first year in office when things were unraveling the Republican leaders in Congress refused to rein him in; indeed, some of them like Representative Devin Nunes appeared to serve as "Trump's stooge" although he was leading the House Committee investigating the Russian interference with the election of 2016.

Free people must exercise liberty or they will see it evaporate in thin air. The degrading of public discourse precedes the degrading of public trust and thrusts the nation into existential travesty. Therefore, all conscious citizens should step forward to declare themselves defenders of the democratic compact that makes us Americans. The only way that we know that we are free is when we are able to make such declarations without duress.

Let us be clear about the real meaning of Donald Trump, his origin and support base beyond what is obvious and conceded. Trump did not come into being by himself or remain in place by himself. He does not live, thrive and spew anti-Mexican and anti-Muslim toxic lines without active and latent supporters. Many of the marchers shouted *weise Menschen uber alles* (White people over all) and "Jews will not replace Us!"

Our response must condemn the vulgar bigotry that defies humanity and makes this nation a repository of unfreedom. Our act of goodwill must begin with us renouncing those who count blood more important than community. Therefore, we cannot sanction any form of demagoguery and we must not open the door to anti-African, anti-Muslim, anti-Semitic, and homophobic rhetoric. We must be antiracist, anti-corporatist, and anti-sexist to regain our pace toward genuine liberation. Let us reject all political stupidity and low-life rhetoric that create insanity in our society. Acceptance of the humanity of the other is the definition of our own humanity; otherwise we are truly mad. Who can explain why racism is morally correct? Who will pursue the mean-spirited demagogues and chase them out of the temples? It can only be us; we are the ones we are looking for and once we accept the challenge the demagogues will have met their match.

Only when we see that values are more important than votes and that being good people is better than being rich people can we overcome the weakness that cause us to believe, at times, that because someone has money he or she also has goodness. What I learned from the people's election of Donald Trump with Russia's assistance is that the fragility of our society is only hanging by the thread of our ethical imagination. Without ethics we are no more than odd political pawns in the game of elections, however empty, and however frequent they are held. To assert ourselves as champions of authentic community means that we will undergo criticism but we will resist Trump and other demagogues who seek to steal our liberation. No single-issue litmus test candidate should be allowed to "trump" profound discourses about the nature of the good. Trump will always support inordinate wealth grabbing, bully domination, deprivation of healthcare and the power of the gun industry parading as the NRA membership. On the other hand, we are stronger than evil and more resilient than the last grand demagogue who thinks that his messianic vision is the one that we should sign onto, but alas, we are merely tough, unadulterated warriors for truth.

Endnotes

1. I was fortunate to have met Casmir during his most productive days at Pepperdine in the 1960s. He was literally a famous professor on campus who taught the only course in communication on persuasion. His recent knowledge of the tactics of Goebbels during the Nazi regime was exceptional since he had participated in that campaign.
2. https://www.facebook.com/events/776196122448656/. This sites provides Casmir's lecture.
3. Erika Wenzel Vora, *The Development of Concept Diffusion Models and Their Application to the Diffusion of the Social Concept of Race*. Buffalo: SUNY-Buffalo, 1978.
4. Benjamin Carter Hett, The Death of Democracy: Hitler's Rise to Power and the Downfall of the Weimar Republic. New York: Holt, 2018. Hett explores how Germany's democracy caved in so quickly and led to the collapse and transformation of the nation's institutions under the rhetoric and bombast of the Hitlerian extremist con.
5. Geoffrey Pridham, *Hitler's Rise to Power*. Endeavour Press, 2016.

Pridham demonstrates how Hitler came to power in Bavaria and went on national power because he was able to ride the wave of anti-Semitism and bigotry he found in the German people. As the country was led to the abyss by euphoria of a pure German state the people lost their freedom and their democracy.

6. *The Washington Post* , January 12, 2018, reported that P resident Donald Trump referred to Haiti and African nations as "shithole countries" during a meeting with a bipartisan group of senators at the White House, a Democratic aide briefed on Thursday's meeting told NBC News.
7. MSNBC, Wolff, Lawrence O'Donnell's Show, January 27, 2018.
8. Ana Monteiro-Ferreira, *The Demise of the Inhuman*. Albany: SUNY Press, 2013.
9. Reinhard H. Luthin, *American Demagogues.* Boston: Beacon Press. 1954, *p. 3)*
10. Robin Wright, "Trump drops the Mother of All Bombs on Afghanistan," *New Yorker*, April 14, 2017.
11. Josh Gerstein, *Politico*, February 15, 2017.
12. Gerstein, *Politico,* February 15, 2017.
13. Aric Jenkins, *Time Magazine,* "Donald Trump's Speech to the NFL on National Anthem Protests," September 25, 2017
14. Caroline Mortimer, "Donald Trump Believes He Has Superior Genes," *Independent*, September 30 2016.
15. *http://www.huffingtonpost.com/entry/donald-trump-eugenics_us_57ec4cc2e4b024a52d2cc7f9*
16. Jack Weatherford, Indian Givers: How the Indians of the Americans Transformed the World. New York: Ballantine, 1989.
17. Weatherford, 1988, p, 128.
18. Adolf Hitler, *Mein Kampf.* Boston: Houghton Mifflin, 1943, pp. 290-293

19. J. Thorley, *Athenian Democracy*. New York: Routledge, 2005; P. J. Rhodes, *A History of the Classical Greek World, 478-323 BC*. New York: Wiley and Sons, 2011.

20. Plutarch, *Pericles*, XIV-XXIII.

21. Donald Kagan, *Pericles of Athens and the Birth of Democracy*. New York: The Free Press, 1991.

22. Daniella Diaz, "Flake Calls Out Trump for Saying Non-Clapping Democrats are Treasonous," *CNN Politics*, February 6, 2018

23. Diaz, *CNN Politics*, February 6, 2018

24. Bogus, Carl T., "The Death of an Honorable Profession" Indiana University Law School, March 17, 1997.

25. These are some of the statements that Trump made over time on the issue. He tweeted on February 18, 2018 "I never said Russia did not meddle in the election, I said "it may be Russia, or China or another country or group, or it may be a 400 pound genius sitting in bed and playing with his computer." The Russian "hoax" was that the Trump campaign colluded with Russia–it never did!" Furthermore, while on Air Force One speaking to reporters on November 2017 Trump said of Putin: "Every time he sees me he says, 'I didn't do that, and I really believe that when he tells me that, he means it." In September 2016 in a debate with Democratic contender Hillary Clinton, Trump said, "She's saying Russia, Russia, Russia, but I don't — maybe it was. I mean, it could be Russia, but it could also be China. It could also be lots of other people. It also could be somebody sitting on their bed that weighs 400 pounds, okay?"

26. Robin Wright writing in the *New Yorker* on January 22, 2017 in an article entitled "Trump's Vainglorious Affront to the CIA," says "In his remarks, Trump made passing reference to the "special wall" behind him but never mentioned the top-secret work or personal sacrifices of intelligence officers like Ames and the others who died in Beirut, including the C.I.A. station chief Kenneth Haas, and James F.

Lewis, who had been a prisoner of war in North Vietnam, and his wife Monique, who was on her first day on the job at the Beirut embassy. Nor did the President refer to any of the dozens of others for whom stars are etched on the hallowed C.I.A. wall of honor. It was like going to the Tomb of the Unknown Soldier and not mentioning those who died in the Second World War."

27. In "The Long and Brutal History of Fake News" *Politico*, December 18, 2016, it was succinctly put that "The fake news hit Trent, Italy, on Easter Sunday, 1475. A 2 ½-year-old child named Simonino had gone missing, and a Franciscan preacher, Bernardino da Feltre, gave a series of sermons claiming that the Jewish community had murdered the child, drained his blood and drunk it to celebrate Passover. The rumors spread fast. Before long da Feltre was claiming that the boy's body had been found in the basement of a Jewish house. In response, the Prince-Bishop of Trent Johannes IV Hinderbach immediately ordered the city's entire Jewish community arrested and tortured."

28. In 1878 Kearney gave a speech that included these words "To add to our misery and despair, a bloated aristocracy has sent to China—the greatest and oldest despotism in the world—for a cheap working slave. It rakes the slums of Asia to find the meanest slave on earth—the Chinese coolie—and imports him here to meet the free American in the Labor market, and still further widen the breach between the rich and the poor, still further to degrade white Labor.

    These cheap slaves fill every place. Their dress is scant and cheap. Their food is rice from China. They hedge twenty in a room, ten by ten. They are wipped curs, abject in docility, mean, contemptible and obedient in all things. They have no wives, children or dependents.

    They are imported by companies, controlled as serfs, worked like slaves, and at last go back to China with all their earnings. They are in every place, they seem to have no sex. Boys work, girls work; it is all alike to them.

    The father of a family is met by them at every turn. Would he get work for himself? Ah! A stout Chinaman does it cheaper. Will he get a place for his oldest boy? He can not. His girl? Why, the Chinaman is in her place too! Every door is closed. He can only go to crime or suicide, his wife and daughter to prostitution, and his boys to hoodlumism and the penitentiary."

29. McCarthy made a distinction between "card-carrying communists" and those who were just "fellow travelers."
30. Holmes, William F. (1970). *The White Chief: James Kimble Vardaman*. Baton Rouge: Louisiana State University Press.
31. Maulana Karenga, *Selections from the Husia*. Los Angeles: University of Sankore, 1989.
32. John McCain wrote an op-ed in the *Washington Post* on January 16, 2017 with the title "Mr. President, stop attacking the press." In the article McCain wrote, "Whether Trump knows it or not, these efforts are being closely watched by foreign leaders who are already using his words as cover as they silence and shutter one of the key pillars of democracy."
33. The entire McCain Op-ed in the *Washington Post* (January 16, 2017) should be read.

CPSIA information can be obtained
at www.ICGtesting.com
Printed in the USA
BVHW04s1524250618
519964BV00030B/1850/P